Emerging Photovoltaic Technologies

Emerging Photovoltaic Technologies

Joel Jean
Swift Solar Inc., San Carlos, CA, USA

Patrick Richard Brown
Massachusetts Institute of Technology, Cambridge, MA, USA

IOP Publishing, Bristol, UK

ISBN   978-0-7503-2152-5 (ebook)
ISBN   978-0-7503-2150-1 (print)
ISBN   978-0-7503-2153-2 (myPrint)
ISBN   978-0-7503-2151-8 (mobi)

DOI   10.1088/978-0-7503-2152-5

Version: 20200801

IOP ebooks

British Library Cataloguing-in-Publication Data: A catalogue record for this book is available from the British Library.

Published by IOP Publishing, wholly owned by The Institute of Physics, London

IOP Publishing, Temple Circus, Temple Way, Bristol, BS1 6HG, UK

US Office: IOP Publishing, Inc., 190 North Independence Mall West, Suite 601, Philadelphia, PA 19106, USA

Joel Jean and Patrick Brown describe the market potential, opportunities, and challenges for emerging solar photovoltaic technologies.

Contents

Preface

Solar photovoltaics (PV) is the fastest-growing energy technology in the world today and an important tool for mitigating climate change. Crystalline silicon PV modules are now affordable, efficient, reliable, and dominant in the global market. So why do researchers and entrepreneurs continue to pursue new PV technologies? This book explores how market forces expose opportunities for new solar technologies. The authors explain how two emerging thin-film PV technologies—metal halide perovskites and colloidal quantum dots—can benefit from rapid scalability, reduced manufacturing and installation costs, and new modes of deployment. This book is targeted at students, early-career researchers, and industry newcomers seeking to maximize their impact in the field of emerging thin-film solar photovoltaics.

Acknowledgements

The authors would like to thank Vladimir Bulović, Tonio Buonassisi, Bob Jaffe, and colleagues from the MIT Future of Solar Study, the Tata-MIT GridEdge Solar program, and Swift Solar for many helpful discussions about the past, present, and future of solar energy. Maximilian Hörantner, David Borrelli, Chi-Sing Ho, and Neal Jean provided valuable feedback on the manuscript.

Author biography

Joel Jean

Joel Jean is a co-founder and CEO of Swift Solar, a US startup developing high-efficiency, lightweight, and flexible perovskite photovoltaics. Joel previously served as the founding executive director of the Tata-MIT GridEdge Solar research program, focusing on scale-up of new solar photovoltaic technologies for India and other developing countries. As a researcher and NSF Fellow at MIT, he developed ultra-lightweight and flexible solar cells that were recognized by the 2017 Katerva Award, and he was named a Forbes 30 Under 30 Fellow in Energy. He co-authored the *MIT Future of Solar Energy Study* and has worked extensively on emerging PV materials and devices, techno-economic analysis, and energy and climate policy. Joel holds a PhD and SM in electrical engineering from MIT and a BS with distinction from Stanford University.

Patrick Richard Brown

Patrick Richard Brown is a postdoctoral researcher at the MIT Energy Initiative. His research focuses on quantifying the near-term competitiveness of photovoltaics, wind, and energy storage across the US electricity system and on technical strategies for integrating high levels of renewable energy into electric power systems. He completed his PhD in physics at MIT, where his thesis research explored the use of colloidal nanocrystals as light-absorbing active materials in thin film solar cells, and his BS in physics and chemistry from the University of Notre Dame. He was a co-author of the *MIT Future of Solar Energy* study and received a graduate certificate in science, technology, and policy for work on the implications of technological advancement in solar and wind technologies for international climate policy.

Chapter 1

The climate challenge and the solar solution

1.1 The big picture: climate change

One of the greatest threats to human civilization is climate change caused by greenhouse gas (GHG) emissions (figure 1.1). From rising sea levels to extreme weather events, the direct impacts of climate change imperil the global economy, human health, agriculture, water supply, and ecosystems. To take just one example, sea level rise and stronger coastal storms increase the risk of storm surge damage and flooding. An average sea level rise of up to 4 feet is expected by 2100, potentially displacing 5 million people and threatening hundreds of billions of dollars of property in the US alone (Melillo *et al* 2014).

Mitigating climate change is the main societal motivation for developing low-carbon energy technologies, such as solar, wind, nuclear, hydro, and ocean power. Over 60% of global GHG emissions—and 75% of all CO_2 emissions—arise from energy use (Baumert *et al* 2005). Our entire energy system must shift from fossil fuels to low-carbon energy—and soon. The longer we wait to act, the greater the cost of mitigation and the risk of severe climate impacts. To keep global warming safely below the 2 °C guardrail, we must reduce GHG emissions by 40%–70% by 2050, which means nearly all electricity generation must come from low-carbon sources (IPCC 2015). This massive decarbonization will likely require well over 10 TW of renewable electricity generation capacity by 2050.

1.2 Why PV: emissions

Solar photovoltaics (PV) is a leading candidate for carbon-free energy generation. It can easily reach the needed terawatt scale without facing severe limits from the available solar resource, land area, and material abundance and use (Jean *et al* 2015). All PV technologies have far lower lifecycle GHG emissions than fossil-fuel generation (<50 gCO_2-eq/kWh for PV versus ~500 for natural gas and ~800 for coal) (figure 1.2). Furthermore, most of the lifecycle emissions of PV today can be

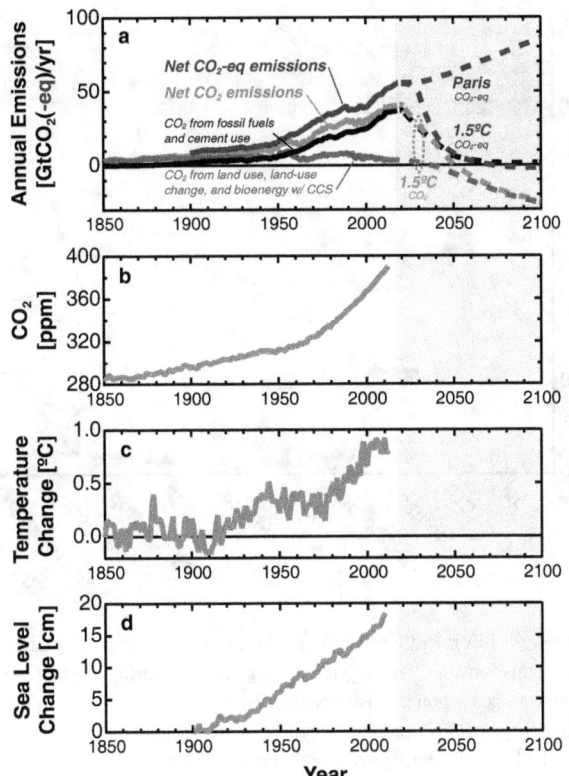

Figure 1.1. Global climate change trends. (a) Annual greenhouse gas emissions, (b) atmospheric CO_2 concentration, (c) global-average land and ocean surface temperature change, and (d) global-average sea level change over time. Panel (a) shows historical CO_2 emissions with separate contributions from fossil fuel and cement use (Boden *et al* 2017) and land use and bioenergy (Meiyappan and Jain 2012), historical total CO_2-equivalent emissions (ClimateInteractive 2017), and two future scenarios—the 2015 Paris Agreement trajectory and a deep decarbonization pathway with roughly a 50% (75%) probability of limiting warming to below 1.5 °C (2 °C) by 2100 (ClimateInteractive 2017, Rockström *et al* 2017). The latter pathway requires a rapid decline in fossil-fuel use and a rapid scale-up of bioenergy with carbon capture and storage (CCS), leading to net zero human CO_2 emissions by 2050. Data are from the IPCC unless otherwise specified (IPCC 2014). Adapted from Jean (2017).

attributed to the grid electricity used in module manufacturing. PV electricity will thus become cleaner as electric power systems worldwide are decarbonized.

By displacing fossil-fuel generation, PV deployment also reduces non-GHG air pollutants. Coal-fired generation, which accounted for roughly 24% of electricity generation in the US in 2019 (EIA 2019) and 39% worldwide (World Bank 2014), emits many pollutants besides CO_2 during operation, including sulfur dioxide (SO_2), nitrogen oxides (NO_x), black carbon, and heavy metals such as mercury and arsenic. SO_2 and NO_x both contribute to the formation of airborne particulate matter, among which PM2.5 (particles with a diameter less than 2.5 μm) is particularly dangerous. The World Health Organization estimates that 4.2 million people die

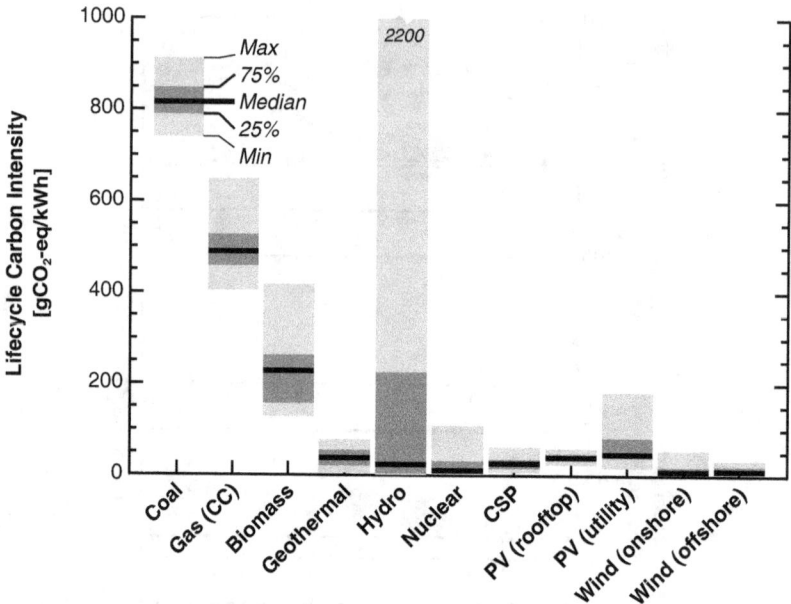

Figure 1.2. Carbon intensity of different electricity generation technologies. The carbon intensity is given in grams of CO_2-equivalent emissions per kWh of electricity generated during the lifetime of a typical generator (IPCC 2015). The global average carbon intensity of electricity was 530 gCO_2-eq/kWh in 2010. Pulverized coal and natural gas combined-cycle (CC) plants, which have traditionally dominated electricity generation, produce more than 10x the life cycle emissions of PV and wind plants.

each year from exposure to outdoor air pollution (WHO 2019). (Coal isn't the only perpetrator—other significant contributors to outdoor air pollution include motor vehicles and, in developing countries, the burning of biomass.)

A number of studies have identified significant benefits to air quality and public health resulting from PV deployment. Millstein *et al* estimate that the solar generation capacity installed across the US between 2007–2015 delivered $2.3 billion in public health benefits, avoiding 300 premature fatalities (Millstein *et al* 2017). (Wind had an even bigger impact over that time period, delivering $29 billion in health benefits and avoiding 6700 premature fatalities.) Continued PV deployment has the potential to deliver much larger benefits: Achieving the US Department of Energy's (DOE) *SunShot* goal of 27% PV energy penetration by 2050 is estimated to lead to ~$167 billion in health benefits, avoiding 25 000–59 000 premature fatalities—even without accounting for the climate benefits of GHG mitigation (Wiser *et al* 2016).

In general, the impact of PV—in terms of economic, health, or climate factors—is dominated by the characteristics of the generation sources that PV replaces. For example, PV modules manufactured in China have higher lifecycle emissions (60–120 gCO_2-eq/kWh) than modules manufactured in the US (30–70 gCO_2-eq/kWh), given China's greater reliance on coal for the electricity used during PV manufacture (Jean *et al* 2015). But this difference is dwarfed by the difference in emissions offset

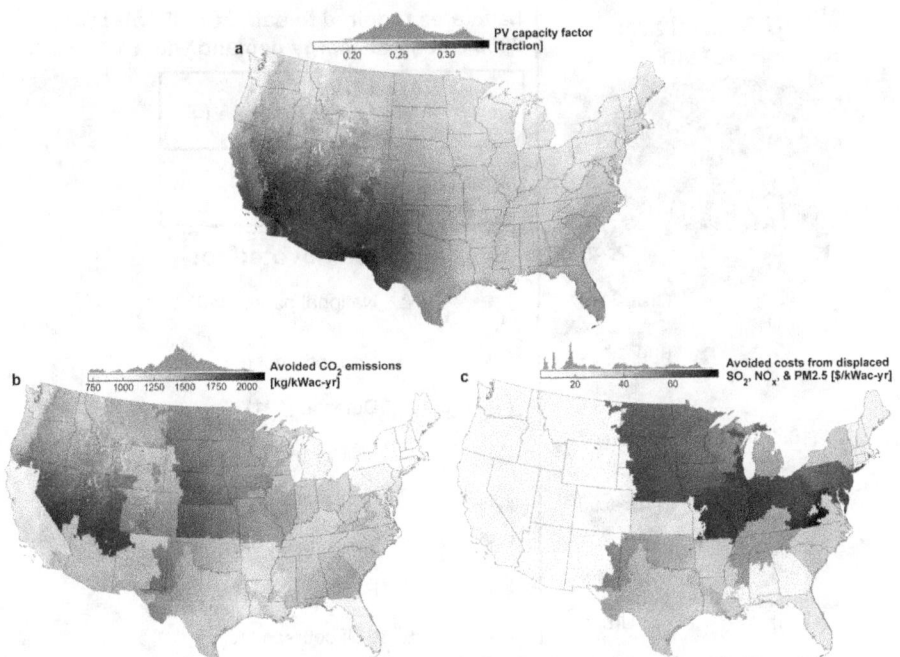

Figure 1.3. Climate and environmental benefits of PV capacity across the US. The hourly AC output of a 1-axis-tracking PV array is simulated at sites across the continental US (a) and multiplied by the marginal emissions rate of CO_2 (b), or the marginal rate of health damages from SO_2, NO_x, and $PM_{2.5}$ (c), from grid electricity in 2017. Marginal emissions rates are quantified by EPA eGRID region, leading to sharp geographic boundaries in (b) and (c). Solar irradiance data are from the NREL National Solar Radiation Database (Sengupta *et al* 2018); marginal emissions data are from Azevedo *et al* (2019). Upstream emissions from methane leakage associated with coal and gas infrastructure are not included.

between the two regions: ~1050 gCO_2-eq/kWh in China, compared with ~600 gCO_2-eq/kWh in the US. Even across the US, differences in local emissions rates across regions lead to differences in the health impacts of PV. Given grid conditions in 2017, the same PV installation would deliver ~6.5× higher health benefits if installed in St. Louis, Missouri, than in Phoenix, Arizona, even accounting for Arizona's much sunnier climate (figure 1.3) (Siler-Evans *et al* 2013).

1.3 Why PV: the solar resource

Even neglecting the climate, environmental, and public health benefits of PV, there are good reasons to expect solar power to contribute significantly to humanity's future energy needs. The scale of the solar resource is immense: enough solar energy strikes the Earth in 2 hours to supply our current energy needs for a year (Schmalensee 2015), and the Sun will supply a stable source of energy for the next five billion years. In the United States, solar PV arrays could generate an amount of electricity equivalent to the entire projected US electricity demand for 2050 on half the land currently used for

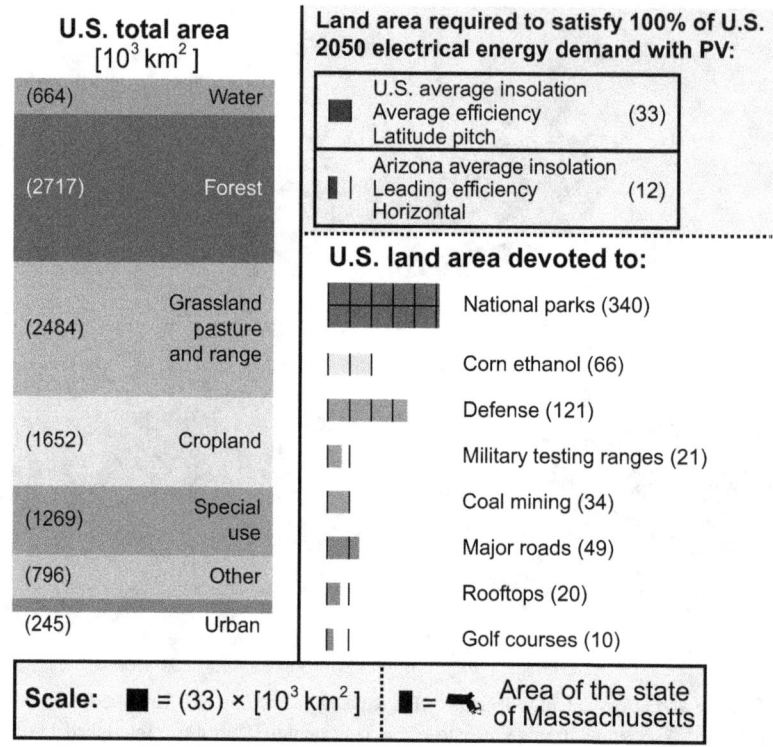

Figure 1.4. Land area required to satisfy 2050 projected US electrical energy demand with PV. Numbers in parentheses denote thousands of square kilometers of area; squares represent the area that would be required for PV, assuming PV meets 100% of demand with no curtailment, storage losses, or transmission losses. Reproduced from Schmalensee *et al* (2015) with permission from the Massachusetts Institute of Technology.

corn ethanol production, or roughly the land area currently taken up by missile testing ranges and golf courses (figure 1.4) (Schmalensee 2015).

At a global scale, solar energy is arguably the most equitably-distributed energy resource. No location in the world lacks direct local access to sunlight. Aggregated over the land area of different nations, annual average solar irradiance varies by at most a factor of four, from ~90 W m^{-2} in Norway to ~360 W m^{-2} in Azerbaijan (figure 1.5). While that difference may sound large—meaning that solar electricity is ~4× as expensive in Norway as in Azerbaijan, all else being equal—this variation in availability pales in comparison with that of other energy resources: the areal density of uranium deposits varies by at least a factor of 1000 between nations, and the density of oil resources by over a million. Solar energy cannot be stockpiled or monopolized, avoiding some of the geopolitical complications associated with relying on concentrated, limited, spatially-heterogeneous resources such as fossil fuels.

We also observe that solar availability is not highly correlated with economic wealth. Figure 1.6 shows insolation and per-capita GDP in 2011 for each country for which these data are available (Jean *et al* 2015). Average insolation is much less

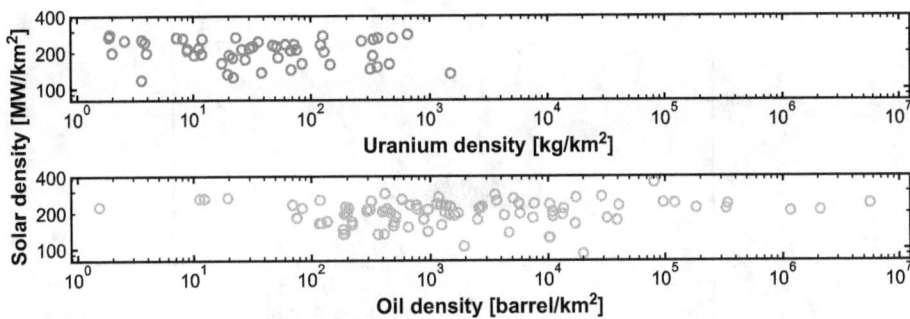

Figure 1.5. Distribution of solar, uranium, and oil densities across nations. Data are from EIA (2019) for oil; OECD (2019) for uranium; UNdata (2019) for area; and OpenEI (2019) for solar illumination. The *x*- and *y*-axes share the same logarithmic scaling. Countries with land area below 5000 km² are not included.

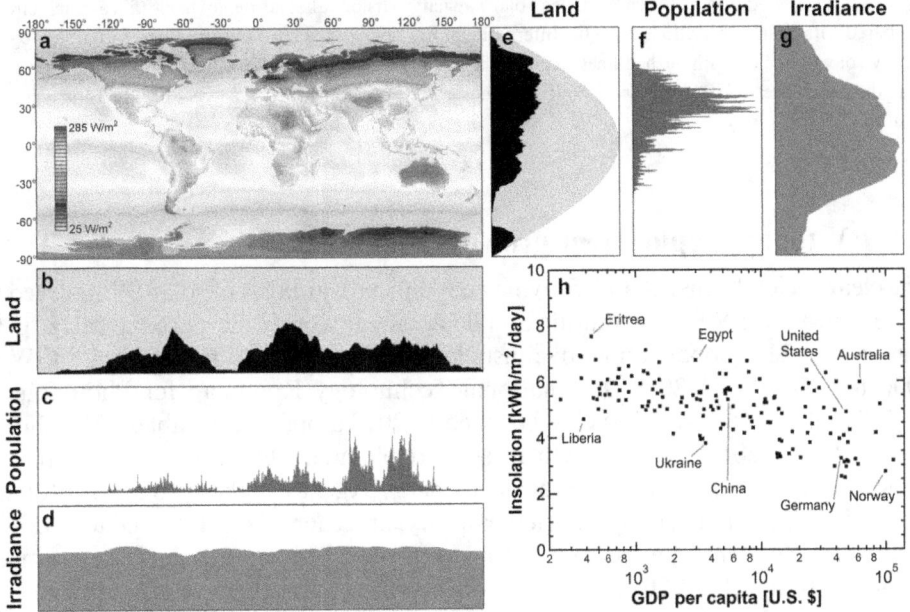

Figure 1.6. Worldwide distribution of the solar resource and correlation with GDP. (a) Global map of solar irradiance [W m⁻²] averaged from 1990 to 2004. Histograms of world land area [m² per °], population [persons per °], and average irradiance at the Earth's surface [W m⁻²] are shown as a function of longitude (b–d) and latitude (e–g). In (b) and (e), land area is shown in black, and water area is shown in blue. (h) Correlation of average insolation and GDP per capita by country for the year 2011. Each dot represents one country. Reproduced from Jean *et al* (2015) with permission from The Royal Society of Chemistry.

variable than GDP, and the weak anti-correlation indicates that developing countries are not fundamentally lacking in access to solar energy. But more importantly, the availability of capital and infrastructure for fully utilizing this resource does vary widely between countries.

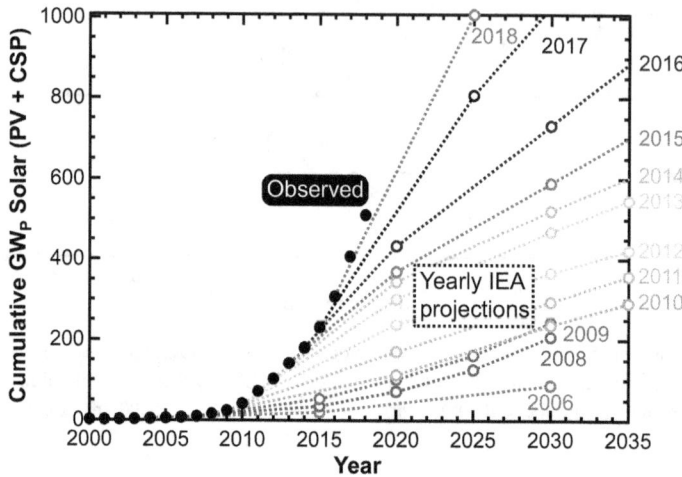

Figure 1.7. Projected and historical installed solar capacity. Historical solar deployment (filled black circles) compared with yearly projections from the International Energy Agency (IEA) World Energy Outlook reports (empty colored circles, with dashed lines to guide the eye). Data include photovoltaics (PV) and concentrated solar thermal power (CSP). Adapted from Schmalensee *et al* (2015) with permission from the Massachusetts Institute of Technology.

1.4 PV today: deployment growth and cost decline

The clear societal benefits and growing economic advantages of solar PV have led to rapid growth in PV manufacturing and deployment worldwide since the early 2000s (figure 1.7). The global cumulative installed PV capacity has grown from 8 GW in 2006 to 500 GW in 2018 (International Technology Roadmap for Photovoltaic, 2017 Results 2018), with 100 GW installed in 2018 alone (Renewables 2019 Global Status Report 2019). This growth in solar deployment has consistently outpaced projections: Each year the International Energy Agency makes projections for the future deployment of solar generation capacity, and each year it is forced to revise its previous projections upward to catch up with observed deployment. For example, according to the IEA's 2014 projection, it should have taken 16 years—until 2030—to reach 500 GW of solar deployment. Instead that capacity was reached in 2018, just 4 years later.

Given the impressive growth in deployment of today's solar technologies (dominated by crystalline silicon), why bother with continued research into new technologies? Is there anything standing between us and a solar-dominated future? In the next chapters we discuss how to think about the competitiveness of solar power, explain the physical and economic hurdles solar has to overcome to supply a larger fraction of our electricity mix, and describe the areas where new PV technologies may be able to out-compete silicon and accelerate PV deployment even further.

References

Azevedo I L, Donti P L, Horner N C, Schivley G, Siler-Evans K and Vaishnav P T 2019 *Electricity Marginal Factor Estimates* (Pittsburgh, PA: Center for Climate and Energy Decision-Making) https://cedmcenter.org/tools-for-cedm/electricity-marginal-factors-estimates/

Baumert K A, Herzog T and Pershing J 2005 *Navigating the numbers: Greenhouse gas data and international climate policy* (Berkeley, CA: World Resources Institute)

Boden T A, Marland G and Andres R J 2017 Global, Regional, and National Fossil-Fuel CO_2 Emissions. Carbon Dioxide Information Analysis Center. https://doi.org/10.3334/CDIAC/00001_V2017

ClimateInteractive 2017 The Climate Scoreboard [WWW Document]

EIA 2019 What is US electricity generation by energy source? https://www.eia.gov/tools/faqs/faq.php?id=427&t=3

International Energy Statistics https://eia.gov/international/data/world (accessed 7.3.19).

International Technology Roadmap for Photovoltaic, 2017 Results 2018 ITRPV

IPCC 2014 Summary for policymakers *Climate Change 2014: Synthesis Report* (Cambridge: Cambridge University Press)

IPCC 2015 *Climate Change 2014: Mitigation of Climate Change: Working Group III Contribution to the IPCC Fifth Assessment Report* (Cambridge: Cambridge University Press)

Jean J, Brown P R, Jaffe R L, Buonassisi T and Bulović V 2015 Pathways for solar photovoltaics *Energy Environ. Sci.* **8** 1200–19

Jean J 2017 Performance limits for colloidal quantum dot and emerging thin-film solar cells *PhD Thesis* Massachusetts Institute of Technology https://dspace.mit.edu/handle/1721.1/111858

Meiyappan P and Jain A K 2012 Three distinct global estimates of historical land-cover change and land-use conversions for over 200 years *Front. Earth Sci.* **6** 122–39

Melillo J M, Richmond T C and Yohe G W 2014 Climate change impacts in the United States: the third national climate assessment *U.S. Global Change Research Program* https://doi.org/10.7930/J0Z31WJ2

Millstein D, Wiser R, Bolinger M and Barbose G 2017 The climate and air-quality benefits of wind and solar power in the United States *Nat. Energy* **2** 17134

OECD 2019 *Uranium 2018 Resources, Production and Demand* (Nuclear Energy Agency) https://doi.org/10.1787/20725310

OpenEI 2019 Solar resources by class and country https://openei.org/datasets/dataset/solar-resources-by-class-and-country

Renewables 2019 Global Status Report REN21 https://www.ren21.net/wp-content/uploads/2019/05/gsr_2019_full_report_en.pdf

Rockström J, Gaffney O, Rogelj J, Meinshausen M, Nakicenovic N and Schellnhuber H J 2017 A roadmap for rapid decarbonization *Science* **355** 1269–71

Schmalensee R 2015 *The Future of Solar Energy: An Interdisciplinary MIT Study* (Cambridge, MA: Energy Initiative, Massachusetts Institute of Technology)

Sengupta M, Xie Y, Lopez A, Habte A, Maclaurin G and Shelby J 2018 The National Solar Radiation Data Base (NSRDB) *Renew. Sustain. Energy Rev.* **89** 51–60

Siler-Evans K, Azevedo I L, Morgan M G and Apt J 2013 Regional variations in the health, environmental, and climate benefits of wind and solar generation *Proc. Natl Acad. Sci. USA* **110** 11768–73

UNdata http://data.un.org/ (accessed 7.3.19)

WHO|Air pollution 2019

Wiser R, Millstein D, Mai T, Macknick J, Carpenter A, Cohen S, Cole W, Frew B and Heath G 2016 The environmental and public health benefits of achieving high penetrations of solar energy in the United States *Energy* **113** 472–86

World Bank 2014 Electricity production from coal sources https://data.worldbank.org/indicator/ EG.ELC.COAL.ZS

Chapter 2

Market drivers for PV research and development

The most basic motivation for continued research into PV technology is to reduce its cost, creating a greater economic incentive to replace fossil-fuelled electricity generation with PV capacity. But how should we think about the cost of PV? What are the right economic metrics for comparing one PV technology against another, and for comparing PV with other electricity generation technologies? Here, we address the different metrics used to quantify PV's cost and value, and discuss the implications of PV's greatest weakness: its temporal intermittency.

2.1 Capacity cost: $/W

The simplest metric for the cost of PV is the upfront cost of PV power generation capacity, measured in dollars per peak watt [$/W]. Peak power output (the 'W' in '$/W') is reported under standardized illumination conditions known as 'AM1.5G', corresponding to an irradiance of 1000 W m^{-2} with a spectrum that closely approximates the spectrum of solar radiation at the Earth's surface (National Renewable Energy Laboratory 2003)[1]. (Most cell testing is performed under calibrated xenon arc or LED lamps rather than under actual solar illumination.)

The '$' in '$/W' can be quantified at a number of different scales: at the level of an individual unencapsulated PV **cell**, an off-the-shelf **module** comprised of multiple interconnected cells, or a complete PV **system** with multiple modules, racking, wiring, and inverters (figure 2.1). Each of these costs can be further broken down into its constituent components. Here, we will use crystalline silicon as an example; for different PV technologies the relative weights of different components may change.

---

[1] The 'AM1.5' refers to an 'air mass' of 1.5, indicating sunlight that has traveled through an equivalent of 1.5 atmospheres, achieved when the Sun is roughly 48.2° from the zenith. The 'G' indicates global irradiance, which includes both direct beam irradiance from the Sun and diffuse (mostly blue) irradiance from the rest of the sky.

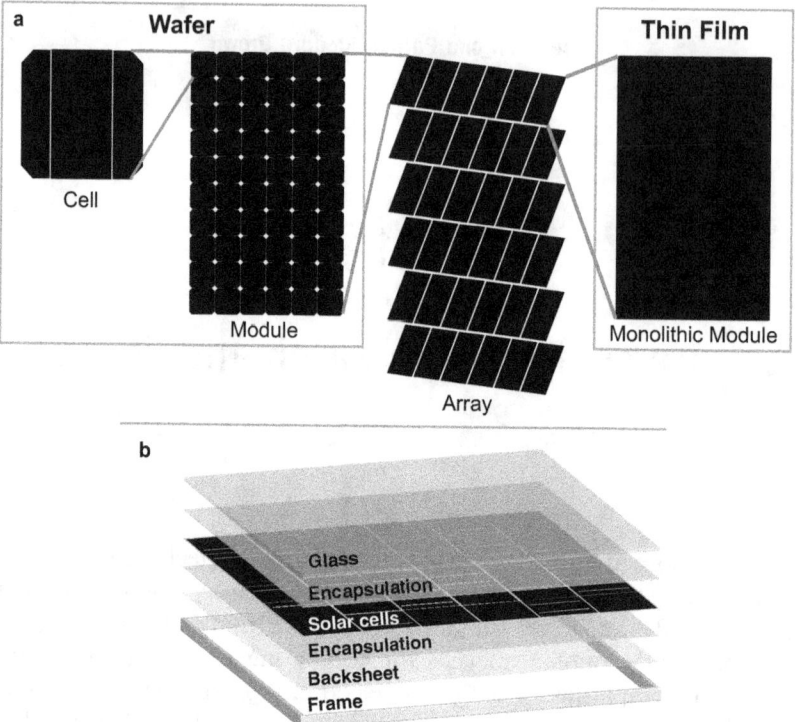

Figure 2.1. Standard solar PV deployment format. Schematics of (a) a grid-connected solar PV system and (b) a typical crystalline silicon PV module. Adapted from Jean *et al* (2015) with permission from the Royal Society of Chemistry.

The PV **cell cost** starts with the bulk cost of the PV active materials: purified silicon for c-Si PV, cadmium and tellurium for CdTe, and so forth. Further costs are associated with the materials used for contacts—including metals (e.g. silver or aluminum for c-Si, molybdenum for CIGS) and transparent conductive oxides (e.g. indium tin oxide, ITO, fluorine-doped tin oxide, FTO, or doped zinc oxide for thin films)—and with the manufacturing processes used to create the PV cell (silicon ingot production, wafering, substrate cleaning, thin-film deposition, doping, texturing, etc).

For crystalline silicon, there are two primary ways to convert polysilicon into silicon wafers ready for cell manufacturing. Monocrystalline or single-crystalline silicon (sc-Si) wafers are made using the Czochralski (Cz) process. Multicrystalline silicon (mc-Si) wafers are made using the directional solidification (DS) process. Both processes involve melting down polysilicon at high temperature, forming it into a solid ingot, and slicing it to produce wafers, although the details of each process step differ substantially. For example, the Cz process produces 150–200 kg cylindrical ingots that are sliced into pseudo-square bricks and wafers, while the DS process forms 800 kg square ingots that are sliced into square bricks and wafers

(Woodhouse *et al* 2019). The final output of both processes is currently a 156.75 mm × 156.75 mm, 150–180 μm thick silicon wafer, although the industry trend is to move toward larger and thinner wafers (International Technology Roadmap for Photovoltaic, 2018 Results 2019). The Cz and DS wafer manufacturing processes are shown in figure 2.2.

Cell manufacturing involves a series of semiconductor and metal film deposition steps and chemical treatments performed on a silicon wafer (for c-Si cells) or a glass, plastic, or metal substrate (for thin-film cells). The exact sequence of steps depends on the cell architecture.

There are two cell architectures and associated processes commonly used for c-Si cell manufacturing. The aluminum back surface field (Al-BSF) process has been an industry standard for many decades for both sc-Si and mc-Si production. The passivated emitter and rear cell (PERC) process has gained substantial market share since 2010 because it can produce higher efficiencies. The PERC process adds a dielectric layer between the silicon wafer and back contact to reduce optical reflection losses and electrical recombination losses at the rear side of the cell. The Al-BSF and PERC processes are shown in figure 2.3.

The **module cost** adds to the cell cost the costs of the window glass, encapsulation and moisture barriers (frontsheet and backsheet), frame, junction box, and costs associated with assembly (i.e. equipment, operating expenses, and labor). An example of the module assembly process for a standard 60-cell crystalline silicon module is shown in figure 2.4. Some of these costs are volumetric and are incurred for each individual PV cell or module (such as the cost of raw materials), while others are fixed capital costs that do not scale with the number of cells or modules produced (such as the cost of manufacturing equipment).

Module cost is more appropriate than cell cost for comparing different PV technologies. This is particularly true for thin-film technologies, as thin-film modules are often manufactured as a monolithic unit rather than assembled from individually-handled cells. As described below, thin-film modules are still separated into cells, but the separation is subtractive rather than additive. Most contemporary PV modules are available at peak capacities of 300–400 W. There is typically a difference between the module manufacturing cost and selling price because the price incorporates indirect costs (e.g. R&D, sales, general, and administrative costs) and a profit margin, depending on market conditions. In recent years, oversupply in the global PV module market has often led to pricing levels close to cost and well below economically sustainable levels (Needleman *et al* 2016). Since 2010, PV manufacturers' average gross margins have ranged between 10%–20%, while average operating margins are typically below 10% and have gone negative in at least one quarter each year (Woodhouse *et al* 2019). These low margins may reflect in part the willingness of Chinese manufacturers to expand capacity by leveraging government subsidies and to overproduce and price products below cost to gain market share.

In the case of silicon, there has been a dramatic and sustained reduction in module cost (and price) over recent decades. The $/W cost of silicon modules has

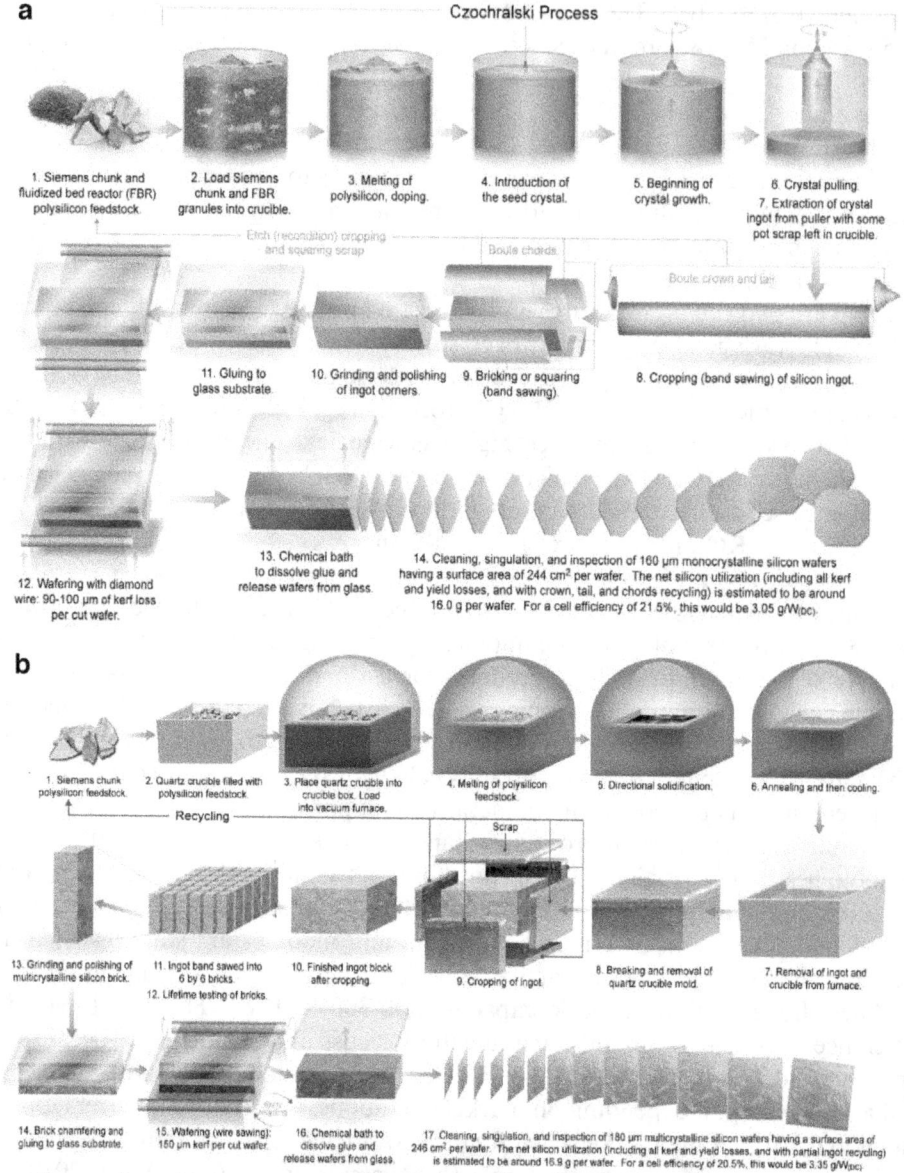

Figure 2.2. Crystalline silicon wafer production processes. (a) Czochralski process flow for producing monocrystalline wafers. (b) DS process flow for producing multicrystalline wafers. Reproduced with permission from Woodhouse *et al* (2019).

fallen by over a factor of 100 since 1980, and by roughly a factor of 10 since 2008 alone (figure 2.5). These cost declines have been driven by simultaneous reductions in bulk silicon costs and in the differentiated manufacturing costs for silicon wafers, cells, and modules, as well as by improvements in cell and module power conversion

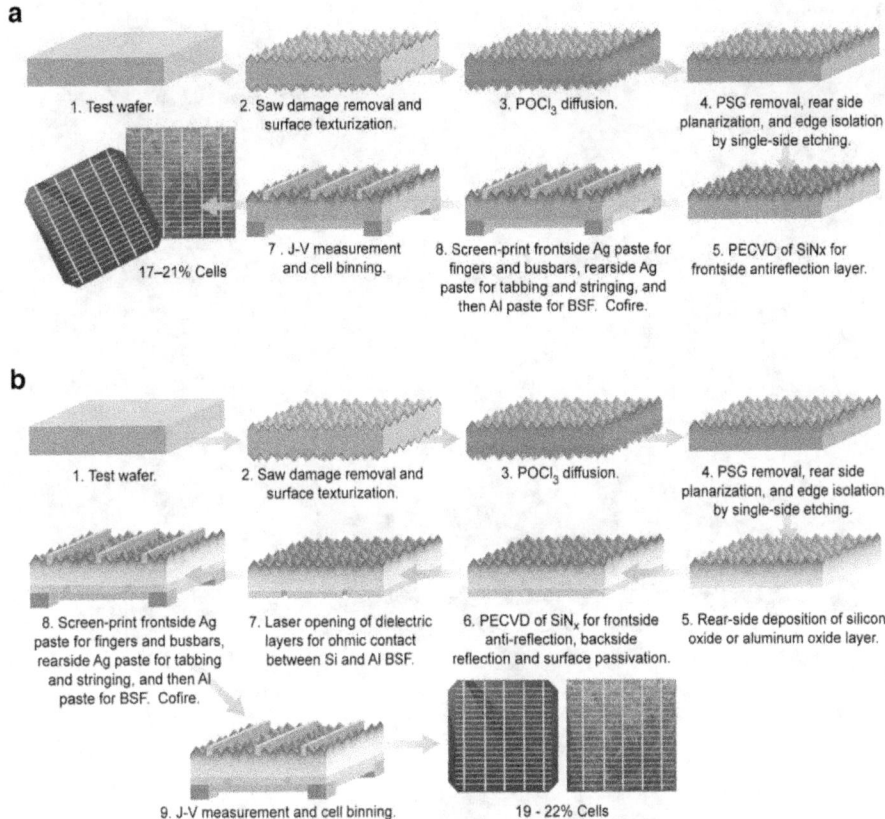

Figure 2.3. Crystalline silicon cell production processes. (a) Industry-standard aluminum back surface field (Al-BSF) process. Key steps include wafer cleaning and texturing, POCl₃ diffusion to produce the active junction, etching of phosphosilicate glass formed after POCl₃ diffusion, plasma-enhanced chemical vapor deposition of hydrogenated silicon nitride (SiN$_x$:H), screen-printing and annealing of Al and Ag pastes for contacts, and cell testing. (b) Higher-efficiency passivated emitter and rear cell (PERC) process. The PERC process adds a rear-side dielectric (SiO₂ or Al₂O₃) deposition (step 5) and laser patterning of the dielectric layers (step 7) to the standard Al-BSF process to achieve higher cell efficiencies. Reproduced with permission from Woodhouse *et al* (2019).

efficiency. At the beginning of 2019, c-Si modules cost less per watt than polysilicon alone in 2010.

The **system cost** adds in the cost of the additional steps necessary to produce useful electricity from a PV module. It incorporates the cost to mount and wire multiple direct-current (DC) modules together (DC hardware costs), convert the modules' native DC output to alternating-current (AC) output using an inverter and transmit that power to the grid or to a local load (AC hardware costs), and actually build the system, which includes labor, permitting, land, financing, and other costs (soft costs) (figure 2.6). These non-module costs are often referred to as 'balance-of-system' (BOS) costs, reflecting both physical hardware ('hard BOS') and labor/administrative costs ('soft BOS').

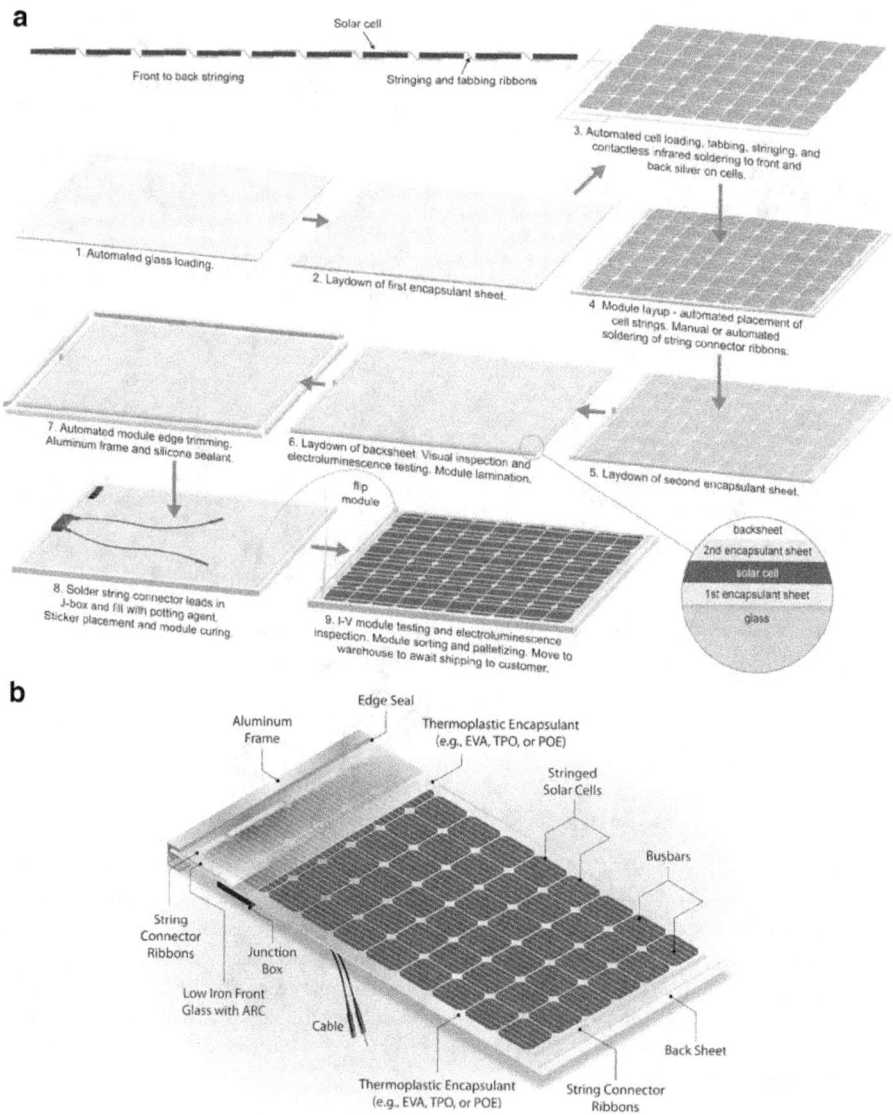

Figure 2.4. Crystalline silicon module assembly. (a) Process flow and (b) finished product for standard 60-cell monocrystalline-silicon module assembly, where J-box is the junction box, ARC is the anti-reflective coating, EVA is ethylene vinyl acetate, TPO is thermoplastic polyolefin, and POE is polyolefin encapsulant. Reproduced with permission from Woodhouse *et al* (2019).

PV systems are usually classified into three market segments: residential (typically <10 kW capacity and mounted on a rooftop), commercial (10–1000 kW capacity, ground-mounted or rooftop-mounted), and utility-scale (⩾1000 kW, ground-mounted and connected to the transmission grid). While residential and commercial arrays consume some of the PV electricity produced on-site and often sell the unused portion back to the utility, utility-scale systems send all of their output through the

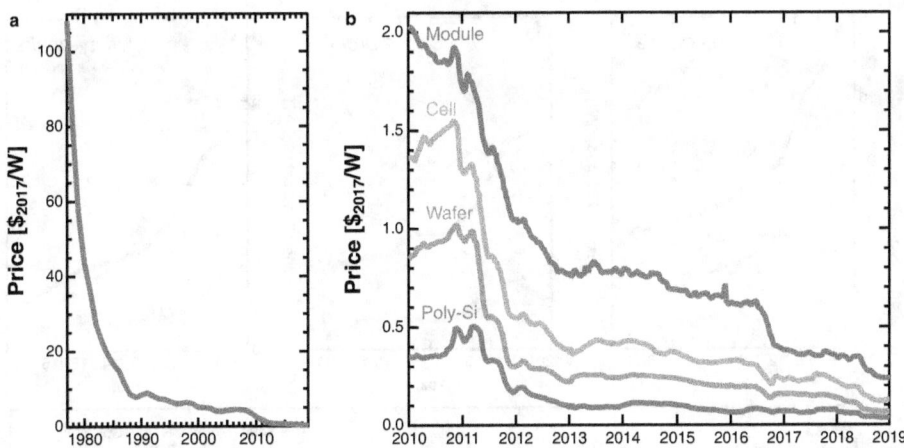

Figure 2.5. Historical PV module price trend and price breakdown by component. (a) Global average c-Si module price and (b) module price decomposition into silicon, wafer, cell, and module assembly costs (International Technology Roadmap for Photovoltaic, 2012 Results 2013, International Technology Roadmap for Photovoltaic, 2018 Results 2019). Prices often vary in different markets due to tariffs and supply–demand imbalances.

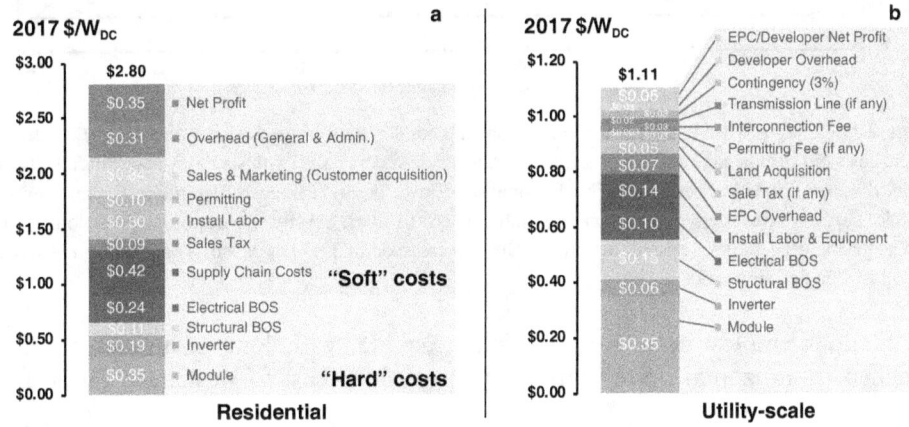

Figure 2.6. System cost breakdown for U.S. residential (a) and utility-scale (b) PV systems in 2017. Note the change in y-axis scale between the two panels. For residential systems (here represented by a 5.7 kW rooftop system), PV modules account for 12% of the system cost and soft costs for 68%; for utility-scale systems (here represented by a 100 MW one-axis tracking system), PV modules account for 31% of the system cost and soft costs for 41%. Data from Fu *et al* (2017).

electric grid, where it is sold either via a long-term contract with a designated off-taker or on the wholesale electricity market. Most utility-scale PV systems also employ **solar tracking** to maximize the irradiance captured by each PV module (Bolinger *et al* 2017). While some arrays use two-axis tracking to align both the module tilt (angle from horizontal) and azimuth (angle from north), most

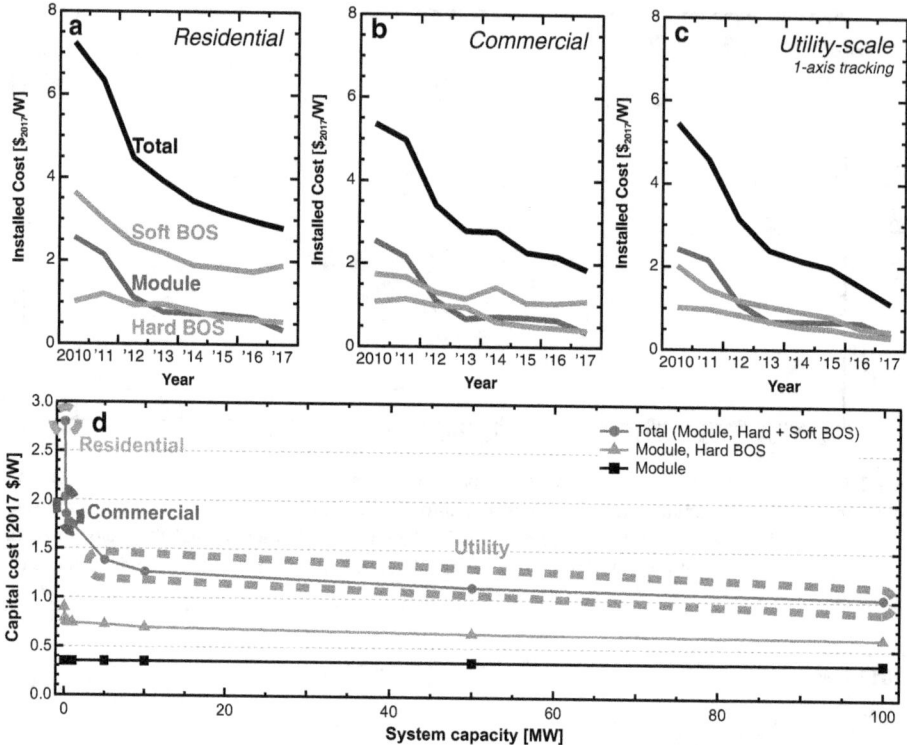

Figure 2.7. Falling PV module, balance-of-systems, and total system costs. Recent trends in (a) residential system costs, (b) commercial system costs, and (c) utility-scale system costs (with single-axis tracking). Data are modeled values for US systems from NREL analysis (Fu *et al* 2017). Since 2010, all cost components— module, hardware (inverter, wiring, racking), and soft costs (labor, permitting, inspection, interconnection, land acquisition, sales tax, overhead, net profit)—have decreased. (d) 2017 capital costs as a function of system size, illustrating economies of scale.

installations employ horizontal one-axis tracking, with long strings of modules attached to horizontal drive shafts that rotate from east to west throughout each day.

While the same PV module can more or less be installed at any scale—from a 12-module rooftop array to a million-module utility-scale array—larger systems tend to have lower BOS costs due to economies of scale, which apply for systems up to ~100 MW (figure 2.7) (Bolinger *et al* 2017). Although residential and commercial systems have been in widespread use since the 1990s, utility-scale PV systems have been deployed widely only since the mid-2000s.

For the PV technologist, capacity cost ($/W) seems like a convenient metric. To a first approximation it is independent of the location of installation[2], and since it is

---

[2] At the system level, some cost components—including the cost of land, labor, permitting, financing, and other factors—can vary across locations (Rhodes *et al* 2017).

quantified under standardized test conditions it is straightforward to use it to compare the costs of different PV module technologies.

However, the location- and application-invariance of capacity cost also means that a PV array installed at the South Pole would have the same capacity cost in $/W as an array at the equator, even though the array at the equator would generate many times more electricity. The capacity cost also does not capture the effects of module lifetime or durability. Two modules may share the same $/W, but if one has a one-year lifetime and the other a 30-year lifetime, the cost of providing a steady source of electricity with these two technologies would be vastly different. PV technologies are also characterized by varying temperature coefficients (representing the loss in efficiency as the cell temperature increases beyond standardized test conditions of 25 °C) and spectral responses, neither of which are captured by a capacity cost measured under AM1.5G conditions. And since a PV module in the field is rarely exposed to illumination at AM1.5G intensities, it is difficult to directly compare PV technologies with other power sources on the basis of capacity cost alone. For these reasons we turn to a more nuanced metric: the **levelized cost of electricity (LCOE)**.

2.2 Levelized cost of electricity (LCOE): $/MWh

In simple terms, the LCOE of any electricity source represents the average cost of producing energy over the lifetime of the generator, in units of $/MWh or ¢/kWh[3]. While the metric of LCOE has a number of limitations (Baker *et al* 2013, Borenstein 2012, Joskow 2011, Reichelstein and Yorston 2013), as will be described below, it remains the most widely-used metric for the cost-competitiveness of solar PV generation.

The LCOE can be calculated using either a complete financial model for a PV generation facility or a simplified model incorporating basic system cost and performance parameters. The financial model approach involves solving for the per-kWh revenue required to achieve a desired internal rate of return. This approach can capture the complex financing arrangements, tax incentives, and electricity market subtleties required for the operation of a utility-scale power plant[4]. However, a full financial model tends to obscure the relationship between techno-economic parameters and LCOE.

A simplified model can clarify the key factors that contribute to LCOE. The LCOE of a PV system can be defined simply as the net present value of all

---

[3] More precisely, the LCOE is the minimum real electricity price that a power plant must receive to break even on investment costs over the life cycle of the facility—in other words, the revenue per kWh needed to achieve a zero net present value over the facility's useful life. This metric accounts for all physical assets and resources required to produce one unit of electricity, including plant capital expenses, cost of capital, and operating expenses.

[4] The System Advisor Model (SAM) (https://sam.nrel.gov/), developed by the US National Renewable Energy Laboratory (NREL), is a widely used financial model that simulates the performance and financial metrics of renewable energy systems in detail. SAM can evaluate the LCOE and other performance metrics for a wide range of PV system configurations in different locations.

construction and operation costs divided by the discounted energy output of the system over its operational lifetime (Baker *et al* 2013, Borenstein 2012):

$$\text{LCOE} = \frac{\sum_0^L C_t/(1 + i)^t}{\sum_1^L E_t/(1 + i)^t} \qquad (2.1)$$

where L is the effective system and financial analysis lifetime (years), i is the real cost of capital or discount rate (%), C_t is the installation and operating costs incurred in year t, and E_t is the energy output in year t. For PV systems, the upfront cost (C_0) dominates the total lifetime cost (numerator). The annual energy output depends on the total nameplate capacity, panel orientation with or without solar tracking, module and system performance losses, shading losses, and local insolation, which depends on the latitude and local climate (e.g. cloud cover, temperature, and humidity) (Jean *et al* 2015).

Typical LCOE ranges for PV and other electric generation technologies are shown in figure 2.8. It is difficult to overstate the magnitude of the shift that has taken place over the last decade: in 2010, PV had the highest LCOE of all widely-deployed generation technologies; in 2019, depending on location, utility-scale PV has the lowest or second-lowest LCOE (second only to onshore wind generation). The era of 'grid parity'—when the cost of obtaining electricity from PV is less than the cost of obtaining it from the grid—seems to have arrived.

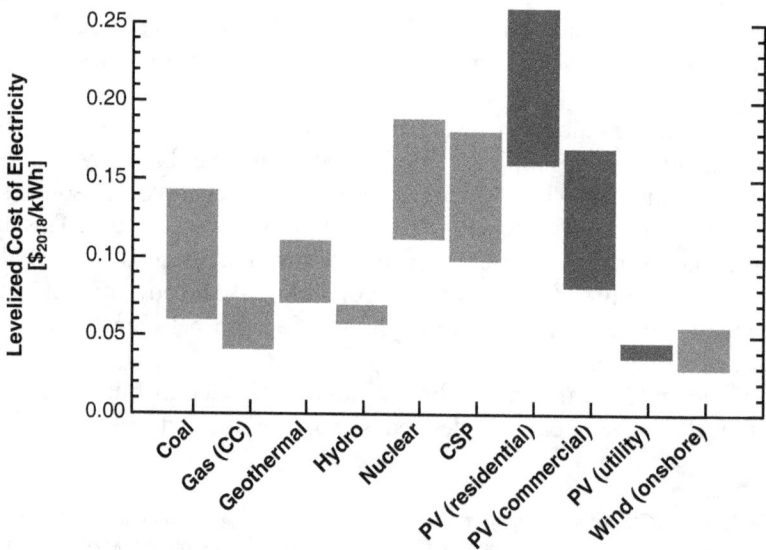

Figure 2.8. LCOE of different electricity generation technologies. Data are from EIA (hydro and offshore wind) and Lazard (all other technologies) (EIA 2017a, Lazard 2018). Unsubsidized LCOE values for US-based systems are shown in units of 2018 USD per kWh. Utility-scale solar PV plants are among the lowest-cost generation technologies available today. Rooftop residential and commercial PV systems are substantially more expensive but have been deployed widely because of government policy, potential indirect benefits, and competition with retail electricity rates rather than wholesale rates.

So is solar already 'solved'? Unfortunately, LCOE leaves out a key feature of the solar resource: its temporal intermittency. As discussed below, intermittency complicates the use of LCOE to compare between generation technologies with different temporal profiles, and in the absence of large-scale energy storage or demand response, the use of LCOE becomes increasingly problematic as solar penetration increases.

2.3 The challenge: intermittency

The **intermittency** of the solar resource can be characterized in terms of two components: the predictable **variability** associated with the daily rotation of the Earth and the seasonal revolution of the Earth about the Sun, and **uncertainty** associated with imperfectly-predictable weather patterns. Uncertainty at the scale of minutes to a few hours can mostly be addressed by spreading solar arrays apart geographically, at a length scale larger than the size of clouds or cloud systems (Mills and Wiser 2010). Variability, while easier to plan for, puts a harder limit on the use of PV to meet electricity demand: the power output of a PV array at night is effectively zero, meaning that the LCOE associated with meeting demand during these hours with PV alone is essentially infinite.

It is the job of the power system operator—a regulated vertically-integrated utility in the traditional arrangement still active in some regions, or an independent system operator, regional transmission organization, or transmission system operator in 'deregulated' markets—to ensure that electricity demand and supply are balanced at all times within its coverage area. In the United States, coverage areas may range from a few counties to multiple states. The electricity use of a single residential consumer varies widely over time and sometimes drops to nearly zero, but when averaged over millions of residential consumers and thousands of commercial and industrial consumers, the total electricity demand within a given coverage area varies relatively smoothly. The periods of peak electricity demand typically occur during weather extremes, corresponding with the use of air conditioning during hot summer days and electric heating during cold winter nights. Importantly, in contrast to the availability of wind and solar over short length scales, the system-wide electricity demand never drops to zero; some amount of generation capacity must operate at all hours of the year to meet demand. Historically, the generation fleet within a given coverage area would consist of a combination of 'baseload' generators (typically nuclear or coal) that operate almost constantly throughout the year, 'load-following' generators that ramp up and down each day to match demand (typically natural gas combined cycle), and 'peaking' generators that sit idle for ⩾90% of the year and only operate during the hours of highest demand.

Figure 2.9 illustrates the temporal profiles of solar generation, wind generation, and electricity demand within the territory of the California Independent System Operator (CAISO)—about 80% of California and a small part of Nevada—in 2017 (CAISO 2017). The 24-hour cycle of solar availability is obvious along each monthly row; the seasonal variability in the length of a day is apparent along a given daily column. There are some synergies between the availability profiles of solar and wind

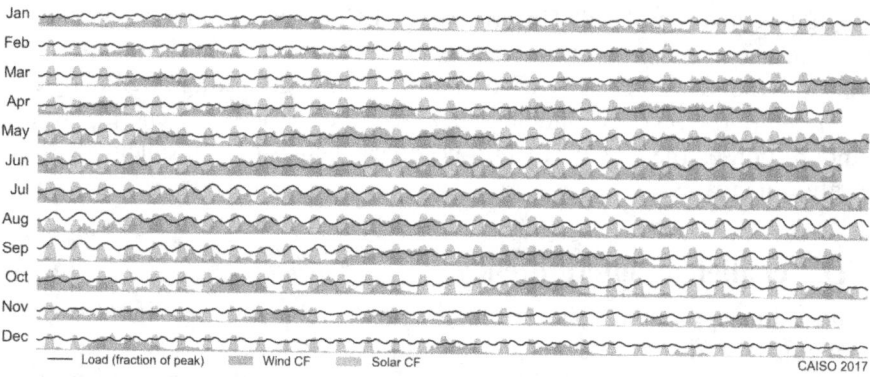

Figure 2.9. Hourly variation in load, solar capacity factor, and wind capacity factor in CAISO 2017. Data are from CAISO (California ISO—Renewables and emissions reports 2017). Load is presented as the fraction of peak 2017 system-wide load; wind and solar are the fraction of peak 2017 wind and solar system-wide generation, respectively.

and the shape of the system-wide load. During the summer months, solar availability is positively correlated with load, as air conditioner usage peaks during the day. Solar and wind also tend to be anticorrelated during the summer months, such that solar availability picks up when wind availability falls and vice versa, presenting a smoother combined availability profile. However, in addition to the nightly outage in solar availability, there are sometimes multi-day periods, particularly in the winter (from January 12–15 in this particular year), when the daily availabilities of both solar and wind are far below their yearly averages.

Some of the seasonal variability in the solar resource can be mitigated through appropriate choice of the PV array orientation. As shown in figure 2.10, the monthly capacity factor (CF)—i.e. the average system output as a fraction of the ideal output under constant peak illumination—of a PV array under typical meteorological year (TMY) weather conditions varies significantly over the course of the year if the array is oriented horizontally: for arrays in El Paso, Kansas City, and Seattle, the July capacity factor is 1.8×, 2.8×, and 7.1× the January capacity factor, respectively. Orienting the array south at latitude tilt mitigates most of this seasonal variability for El Paso and Kansas City (where the new July/January CF ratios are 1.0 and 1.4), though it has less of an effect in Seattle where the winters are particularly cloudy (July/Jan CF = 3.4). Horizontal one-axis east-to-west tracking extends PV generation earlier into the morning and later into the evening, increasing annual capacity factors by 15%–20% over fixed-latitude-tilt systems (though with a small decrease in winter capacity factor and mid-day generation).

However, none of these arrangements can get around the fact that the solar resource is available for no more than half the hours in the year, capping the amount of electricity demand that PV can offset on its own. Figure 2.11 illustrates this effect

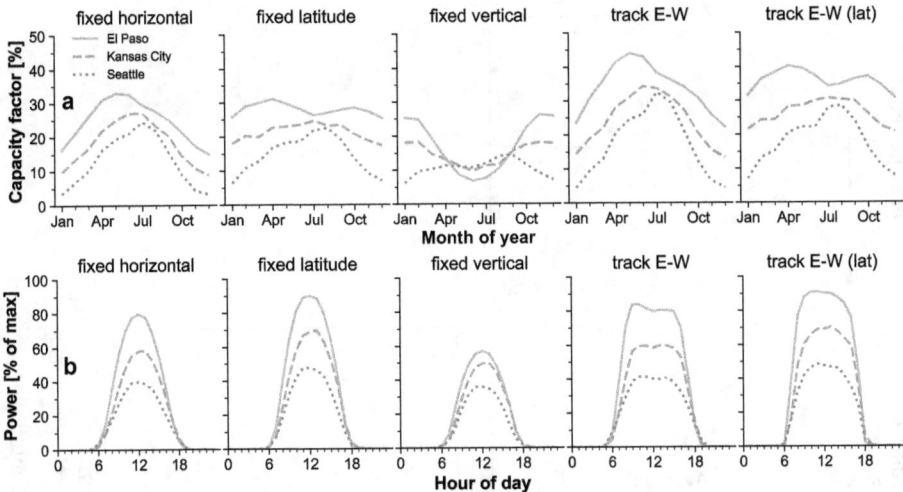

Figure 2.10. Monthly capacity factors for PV arrays in three different US cities: El Paso, TX; Kansas City, MO; and Seattle, WA. Cardinal directions apply to PV generators in the Northern Hemisphere. 'Fixed latitude' refers to south-facing tilt at an angle equal to the latitude; 'fixed vertical' refers to south-facing tilt at 90°; 'track E-W' refers to single-axis tracking from east to west along a horizontal tracking axis (the most common tracking arrangement); 'track E-W (lat)' refers to single-axis tracking from east to west along a tracking axis tilted south at the angle of the latitude. Values correspond to AC power generation as a fraction of peak AC power capacity averaged over a typical meteorological year, grouped by month of year (a) and hour of day (b).

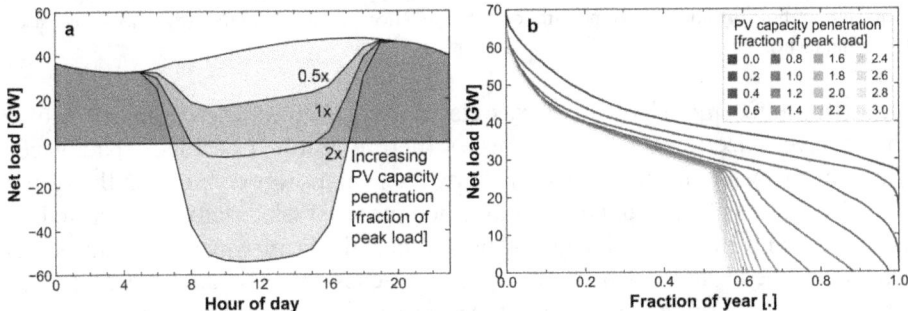

Figure 2.11. Declining returns to PV capacity. (a) Hourly average ERCOT (Texas) load in 2017 minus simulated one-axis-tracking PV generation, where the PV capacity is equal to 0.5× (yellow), 1× (orange), and 2× (red) peak load. Light orange and red areas below the $y = 0$ line represent curtailed PV generation. (b) Simulated hourly net load (load minus PV generation) for the same system as a function of PV capacity penetration, with each hour of the year sorted in order of decreasing net load (a representation known as a 'load duration curve' or, in this case, a 'net load duration curve'). As PV penetration increases, its marginal impact on the net load profile becomes increasingly small, as increasing amounts of PV generation are occurring during hours when PV is already being curtailed.

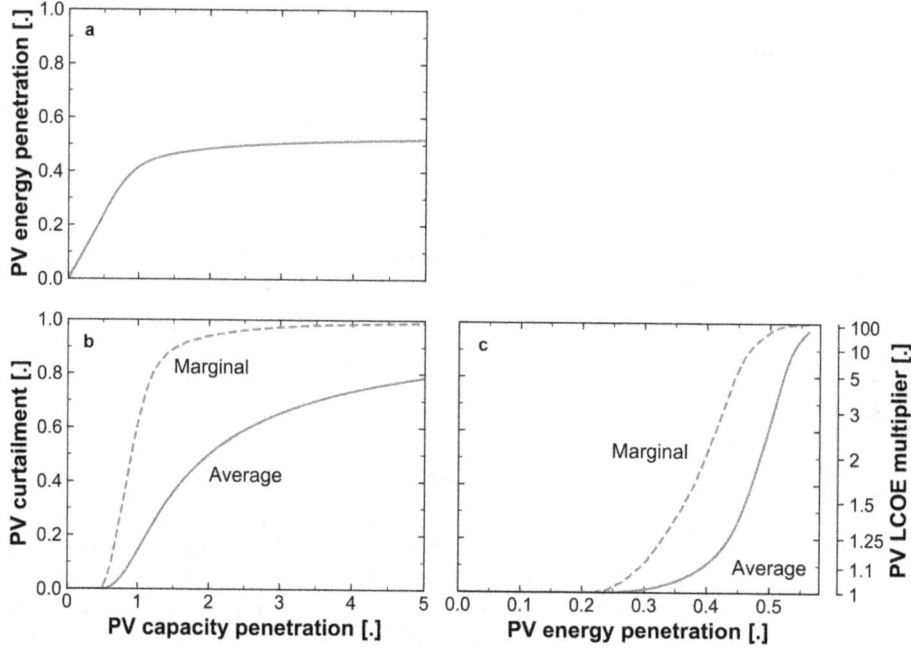

Figure 2.12. PV energy penetration and curtailment. Results are simulated using ERCOT 2017 load and one-axis-tracking PV profiles. (a) PV energy penetration (the fraction of yearly demand met by PV) rapidly saturates after PV capacity penetration (PV power capacity as a fraction of peak load) increases beyond 1. (b, c) Marginal PV curtailment (dashed lines) and average curtailment (solid lines) as a function of PV capacity penetration (b) and energy penetration (c). The effective LCOE of PV (right axis) increases as $1/(1 - \text{curtailment})$.

using a simplified model of the Texas electricity grid[5]. As the deployment of PV within a given electricity system increases (here quantified as the 'capacity penetration', or the peak output of PV divided by the peak power demand on the system), eventually there will be hours of the day when more PV electricity is generated than can be used, and the excess PV generation is 'curtailed', or wasted. On this system, PV begins to be curtailed when PV capacity reaches roughly half of peak load. As PV penetration continues to increase, the number of hours with zero 'net load' (load minus PV generation) increases, and increasing amounts of PV generation are curtailed in an effort to supply the remaining hours in the morning and evening (compare the size of the light and dark red areas in figure 2.11(a)).

The increase in curtailment with PV penetration has two important effects. First, it puts a limit on the 'energy penetration' that PV can achieve—the fraction of total yearly electricity demand that can be met by PV (figure 2.12) (Denholm and Margolis 2007a). In the absence of curtailment, energy penetration increases linearly with capacity penetration; once curtailment begins to occur, each unit of additional

---

[5] In this simplified representation, we ignore the influence of a number of effects for non-PV generators, including minimum generation levels, minimum uptime and downtime durations, and ramping limits, and assume that demand is price-inelastic (i.e. the hourly demand is fixed).

PV capacity supplies a diminishing amount of electricity demand, and the rate of increase of energy penetration slows (figure 2.12(a)). At an energy penetration around 40%, the 'marginal curtailment' rate—the curtailment rate of the next unit of PV capacity to be added to the system—reaches 50%, and the energy penetration begins to saturate[6]. By the time PV energy penetration reaches 47%, roughly 90% of the generation of new PV capacity is being curtailed.

Second, the curtailment rate has a direct impact on the LCOE of PV. According to equation 1, the LCOE is inversely proportional to the energy delivered over the plant's lifetime—so as curtailment increases and delivered energy decreases, LCOE increases as 1/(1–curtailment). For the 40% and 47% energy penetration cases noted above (with marginal curtailment rates of 50% and 90%), the effective LCOE of new PV generation is 2× and 10× the base-rate LCOE. Achieving grid-parity at low rates of PV penetration is thus very different from achieving it at 40% or 50% PV penetration.

Even before curtailment becomes a significant issue, PV's competition begins to get tougher as PV penetration increases. A utility-scale PV generator participates in the wholesale electricity market and gets paid a time-varying rate in $/MWh for the electricity it produces. The different generators in a given power system are dispatched according to their 'merit order': that is, in order of their marginal generation costs, with the cheapest generators dispatched first. The last and most expensive generator that is dispatched to meet demand in a given hour (the 'marginal generator') sets the electricity price at its marginal $/MWh cost, and the rest of the generators that are dispatched in that hour are paid this same $/MWh rate. Conversely, this is the rate paid by large consumers to consume electricity during this hour. (The demand from residential and small commercial customers is aggregated by distribution companies; these companies buy electricity on the wholesale market at the time-varying rate, but often charge their customers a fixed retail price for electricity regardless of when it is used. Prices on the distribution system include additional markups to cover the cost and maintenance of distribution lines, efficiency and renewable energy programs, and distribution company profits.)

Figure 2.13 shows a hypothetical merit-order curve, with renewables at the bottom (since they require no fuel, their marginal costs are almost zero), followed by nuclear, coal, natural gas combined cycle, and natural gas and petroleum peakers. (Depending on the relative prices of coal and gas, the order of generators in the merit order may vary.) Thus periods with high electricity demand tend to have a high electricity price. The very highest demand hours in the year can witness particularly high price spikes, as older, less-efficient, and more-expensive plants are dispatched to meet the peak demand.

---

[6] We here assume that the energy from the first units of PV to be installed is utilized first, and thus is not curtailed as new PV capacity is added. This assumption approximates the behavior of first-come first-served contracts between PV generators and the utility. Thus the *average* fleet-wide curtailment rate across all PV generators in the system (solid lines in figure 2.12) is lower than the marginal curtailment rate. If all PV generators are instead uniformly curtailed, the marginal curtailment would equal the average curtailment.

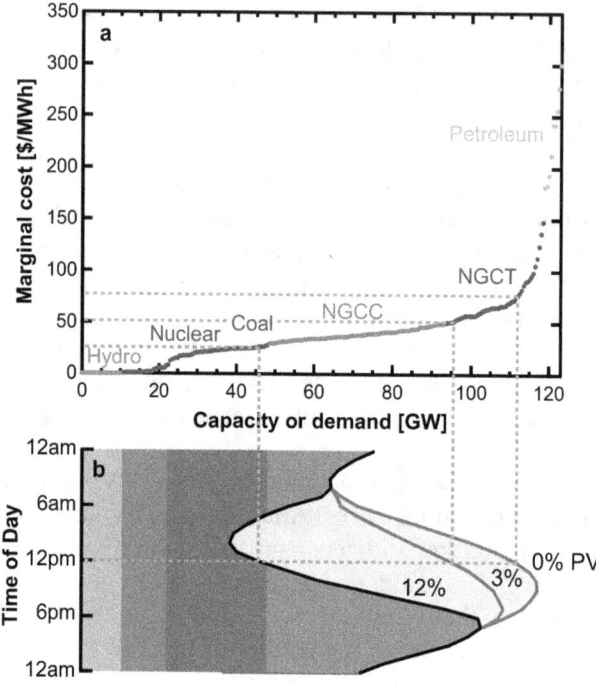

Figure 2.13. Merit-order effect. Generators are dispatched in increasing order of their marginal generation costs (a); the most expensive generator dispatched in a given hour sets the price that all generators are paid. As PV penetration increases (b), the mid-day electricity price decreases and all PV generators are paid less for the electricity they produce. 'NGCC' denotes natural gas combined cycle; 'NGCT' denotes natural gas combustion turbines. The hypothetical merit-order curve in (a) is adapted from EIA (2012).

As noted above, solar availability tends to be correlated with demand, so the first few PV generators on a system enjoy a high electricity price for their generation. This price is even higher than the average electricity price over the year: if the average price is 50 $/MWh, a solar generator may make 55 $/MWh on average since it tends to be most available at high-demand, high-price times. But as solar generation increases, the most expensive plants are displaced first, and the electricity price begins to fall during precisely the hours when solar is generating (Hirth 2013, Mills and Wiser 2012). This effect is already being observed in California (figure 2.14): Before 2012, electricity prices tended to be at a maximum in the middle of the day, while in 2017, prices were at a minimum during the middle of the day (when solar generation is strongest).

The experience thus far with relatively high PV penetrations in California and other regions has shown that electricity prices don't just decrease as PV generation increases; they sometimes go negative. Negative prices arise from either of two unrelated phenomena. First, some inflexible generators including large coal and nuclear plants incur significant maintenance requirements, fuel costs, and operational delays from repeated start-up and shutdown events, resulting from thermal ramp limits on large steam boilers for both generator types and from undesirable

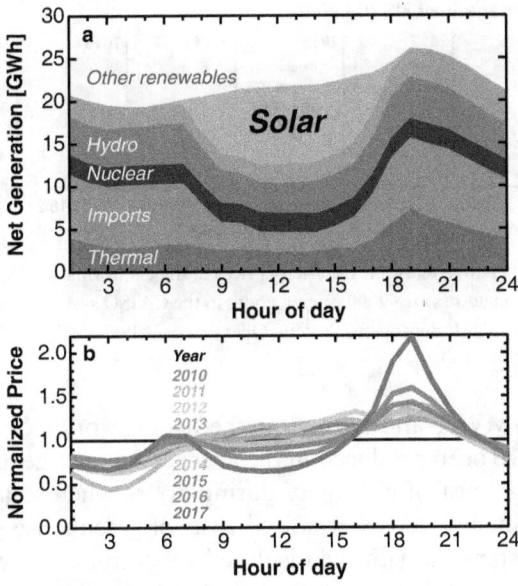

Figure 2.14. Effect of high solar penetration on wholesale electricity market prices. (a) Net hourly electricity generation in California on March 11, 2017 (EIA 2017b). Solar PV now accounts for over 10% of all electricity production in CA. (b) Day-ahead locational marginal prices—roughly the local wholesale electricity price with transmission constraints—averaged over all nodes in California for different years and normalized to the annual-average price. High solar generation reduces the mid-day electricity price and thus the remuneration received by PV system owners. Price data are from the California Independent System Operator (CAISO 2017).

decay products that occur during ramping of nuclear generators (Jenkins *et al* 2018). During short periods of low demand, these generators thus bid negative prices into the wholesale market to ensure they are continually dispatched, effectively paying consumers to use their energy. Second, some subsidies for renewable energy, including state renewable energy credits and the federal production tax credit for wind, are applied on a per-MWh-generated basis. A generator receiving a 5 $/MWh out-of-market subsidy would therefore be willing to be dispatched even if prices fall to −5 $/MWh, and would thus bid −5 $/MWh as its marginal cost. While PV does not directly cause negative prices, adding PV to a system with a large amount of inflexible generation can lead to an increased incidence of negative prices during the hours when PV is generating, as has been observed from 2014–2017 in CAISO (figure 2.15). Negative prices are good news for consumers and bad news for coal, but they also lead to curtailment of PV above and beyond the demand-limited curtailment discussed in figure 2.12, further increasing the effective LCOE of PV.

2.4 The value of solar electricity

In addition to thinking about the cost of solar, we therefore need to think about the *value* of solar electricity. One metric for the value of solar is the levelized avoided cost of electricity (LACE) (US Energy Information Administration 2013). LACE is

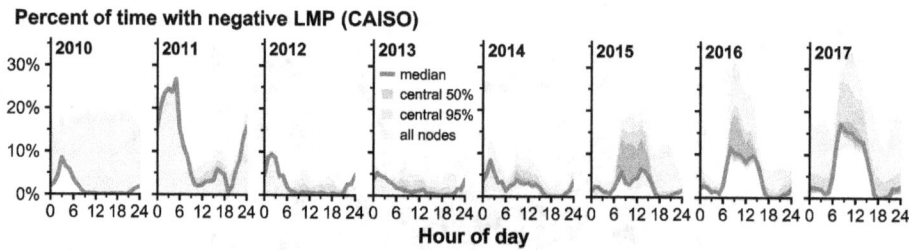

Figure 2.15. Incidence of negative real-time electricity prices in the California ISO (CAISO) over time. Values are shown as the distribution across all ~2000 pricing nodes in the CAISO system, binned by hour of day. Price data are from the California Independent System Operator (CAISO 2017). Adapted from Brown and O'Sullivan (2019).

also measured in $/MWh, and represents the average price of the electricity that would otherwise have been produced in the absence of solar generation—effectively the weighted average cost of electricity during hours when solar is available. The true LACE should include the externalized costs of greenhouse gas and particulate matter emissions associated with the displaced generators, as well as avoided costs from the capacity market (a marketplace used in some but not all jurisdictions to reimburse generators for being available to meet demand during times of peak load) and the ancillary service market (a marketplace that reimburses generators for providing standby reserves and ramping capacity to respond to unforeseen changes in supply and demand).

Solar can be said to be competitive at a given penetration when its LCOE is below the LACE. Given that the LACE will continuously decline as solar penetration increases, the cost of solar needs to continue to fall in order to stay competitive at higher levels of deployment. (In a similar effect, even the environmental and climate value of solar tends to decline as solar penetration increases, as the most-expensive plants that solar displaces first also tend to be the dirtiest (Millstein *et al* 2017).)

The LCOE metric must be used with caution, as it does not account for the temporal coincidence between a given energy resource and the demand for electricity or the availability of other resources. LCOE is not generally suitable for comparing the cost of intermittent PV at high penetration with other non-intermittent generators. Nonetheless, there are some specific cases where LCOE is still a valid metric for comparison:

1. Comparing **different PV technologies in the same location and orientation.** If two technologies have the same temporal availability profile, LCOE can safely be used to compare their relative costs. As noted above, differences in temperature coefficient and spectral response between PV technologies can modify their respective temporal output profiles (particularly between seasons), so this method becomes less applicable as these parameters diverge.

2. Comparing PV to other generators when the **cost of the non-PV alternative is uniform in time**, and the amount of PV being added is not enough to saturate demand. This situation applies, for example, in some rural off-grid settings where diesel generation is the only other option for electricity generation, or

    in hydropower-dominated systems where PV generation serves to conserve water resources for later use.

3. Comparing PV to other generators when there is **enough energy storage or flexible demand** to enable the use of all electrical energy produced by PV (or any other source) regardless of when it is produced. As grid-scale electricity storage and demand response are not yet widely deployed, this condition is not met in power systems today, but may apply in the future if dramatic cost reductions for these PV-enabling technologies are achieved.

4. Comparing PV to the **combined cost of alternative generation using the LACE.** The relative competitiveness of a given technology increases as (LACE—LCOE) increases. In this case PV is not compared directly to a specific technology alternative, but to the collection of generators within a given power system that would otherwise supply electricity in the absence of the new PV generator. The LACE must therefore be continuously reassessed as PV deployment increases or other conditions on the power system change.

Unlocking the potential of PV for large-scale energy generation can therefore be achieved in two ways: by **decreasing the LCOE**, thus enabling PV to remain competitive even in the face of curtailment and electricity price decline; and by **increasing the LACE**—that is, by delivering PV electricity at the times and locations where it is most valuable.

There are a number of strategies for increasing the value of PV (Delucchi and Jacobson 2011, Denholm and Margolis 2007b, Mills and Wiser 2015):

Demand response: The hourly profile of electricity demand is not necessarily fixed. To the extent that the time-varying price of electricity is communicated to electricity consumers and those consumers have means of adjusting or scheduling their electricity use, demand can be moved from low-solar-availability (high-price) hours of the day to high-solar-availability (low-price) hours. All sides benefit from this arrangement: consumers get cheaper electricity, and PV generators enjoy increased demand (and therefore prices) during hours when they are generating. A common approach to demand response is to use thermal inertia to temporally decouple the electricity use of heating/cooling systems from the delivery of heating/cooling services: PV electricity can be used during daytime hours to freeze or heat water, which can then deliver cooling or heating services at night. In general, in a solar-dominated future where the marginal price of electricity can be expected to be near zero for 8+ h per day, people will likely find productive ways to use cheap electricity.

Demand growth: Decarbonizing our society will require replacing point-of-use fossil fuel consumption (for applications such as residential heating, cooking, and transportation) with electric alternatives. Electrification of new sectors will result in a large increase in the demand for zero-carbon electricity. Electric vehicles in particular represent a large potential source of new demand that could easily be met by PV if there are adequate incentives and resources for daytime (typically workplace) charging. Climate change is also likely to increase demand for water desalination and air conditioning, both of which are electricity-intensive applications that will be most important in sunny regions.

Power system flexibility: 'Baseload' generators such as nuclear, coal, and natural gas combined cycle (CCGT) plants often have minimum-generation levels—on the order of 50% of peak output, lower for CCGT and higher for nuclear—below which they are not able to operate. To avoid turning off the generator completely and incurring operational costs associated with shut-down, start-up, and minimum-off-time requirements associated with thermal stress limits on large boilers, these baseload generators often bid negative prices to avoid being turned off, as discussed above. Negative prices hurt all generators on the system, including solar, some of which may be curtailed during such periods (though fuel-based generators, which still incur fuel costs during negative-price intervals, are hurt the most). We can help the power system adapt to higher levels of variable generation from PV and wind by refurbishing existing baseload generators to operate more flexibly and ensuring that any newly-installed nuclear capacity uses flexible system designs. Even without adding new infrastructure to the power system, implementing more flexible electricity market structures and improving the accuracy of weather and demand forecasts can ease the integration of variable renewables (Bird *et al* 2013).

Long-distance transmission: Even in relatively sunny California, cloudy periods sometimes persist for many days in a row (figure 2.9). Cloudy periods persist even longer in the northwest and northeast, and the wind resource behaves similarly. However, at the continental scale, it is almost always sunny or windy somewhere (figure 2.16). Installing additional long-distance electricity transmission lines—particularly high-voltage DC lines, which suffer lower losses over long distances than AC lines—can enable increased solar and wind energy penetrations at lower levels of installed PV and wind power capacity (Kempton *et al* 2010, MacDonald *et al* 2016).

Resource complementarity: Figure 2.9 also shows that, in some regions and during some times of the year, the availability of solar and wind can be complementary, with wind generation strongest at night and solar generation occurring during the day. Relying on a combination of PV and wind, as well as hydropower, geothermal, and perhaps nuclear, thus enables a zero-carbon power system to be achieved at lower cost than relying on PV alone (Heide *et al* 2011, Kreifels *et al* 2014, Sepulveda *et al* 2018).

Energy storage: Utility-scale energy storage provides perhaps the most direct solution to PV intermittency; while it is still relatively expensive, prices are dropping quickly, particularly for lithium-ion batteries (Schmidt *et al* 2017). As shown in figure 2.17, while storage is not required and does not significantly impact the ability of PV to meet demand at low levels of solar penetration, at higher penetration levels (beyond ~40% PV energy penetration) storage prevents PV curtailment and raises the upper bound on energy that can be provided by PV. Looking beyond electrical energy storage, chemical storage (in the form of hydrogen, for example) also enables PV energy to be used in applications beyond the electric power system such as transportation or industrial process heat.

While energy storage is necessary for achieving solar energy penetrations above 50%, it is likely to be more economical to 'overbuild' PV capacity and use storage for daily balancing than to attempt to minimize PV curtailment using 'seasonal' storage;

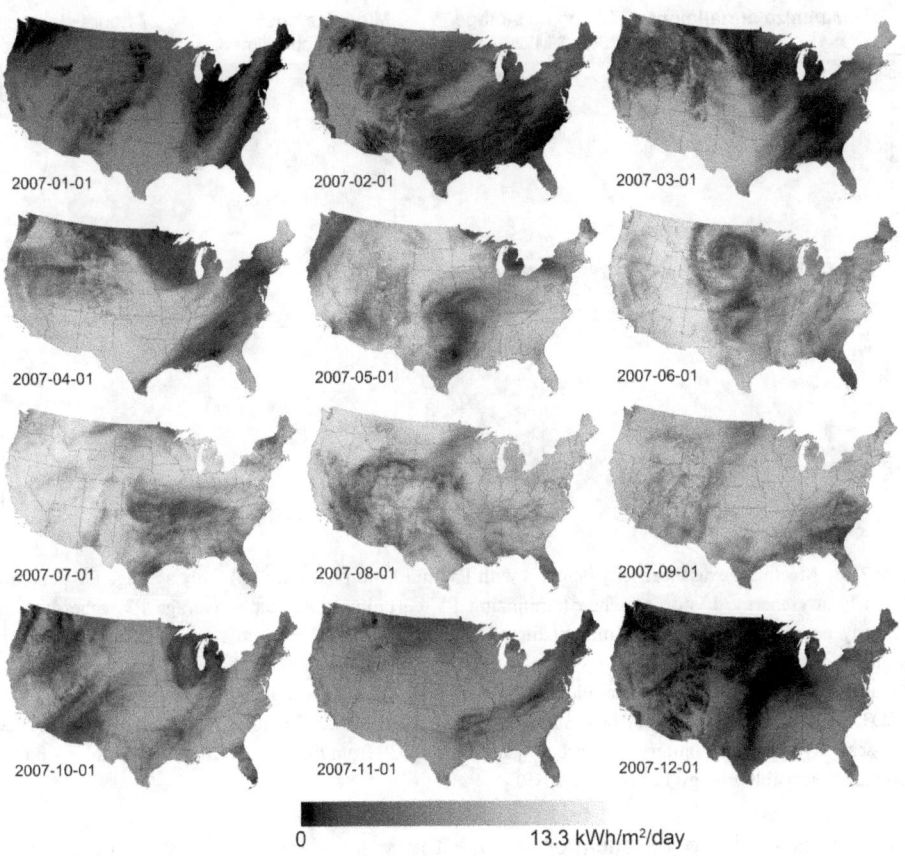

0   13.3 kWh/m²/day

Figure 2.16. Daily average simulated generation of one-axis-tracking PV arrays on the first day of each month of 2007. On some days, large regions of the country are obscured by clouds while other regions enjoy clear skies. In general, the sky is almost always clear somewhere; long-range high-voltage transmission would smooth out this variability and facilitate the use of PV to supply a large fraction of electricity demand. Insolation data are from the National Solar Radiation Database (NSRDB) (Sengupta *et al* 2018).

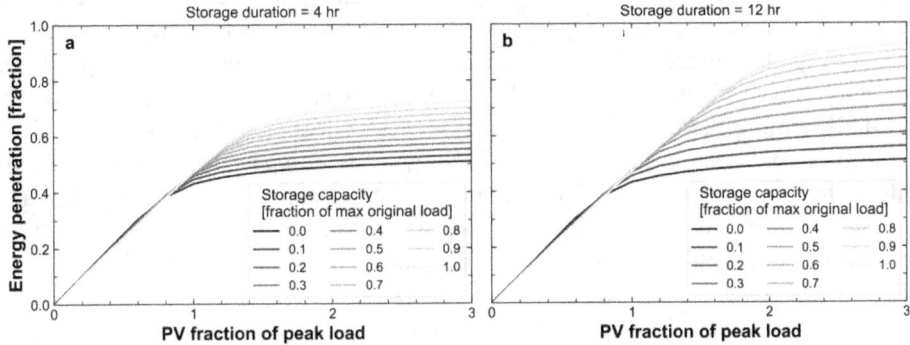

Figure 2.17. Storage increases utilization of PV energy at high PV penetrations. Results are simulated for one-axis-tracking PV on the ERCOT system in 2015 assuming 4 h-duration storage (a) and 12 h-duration storage (b).

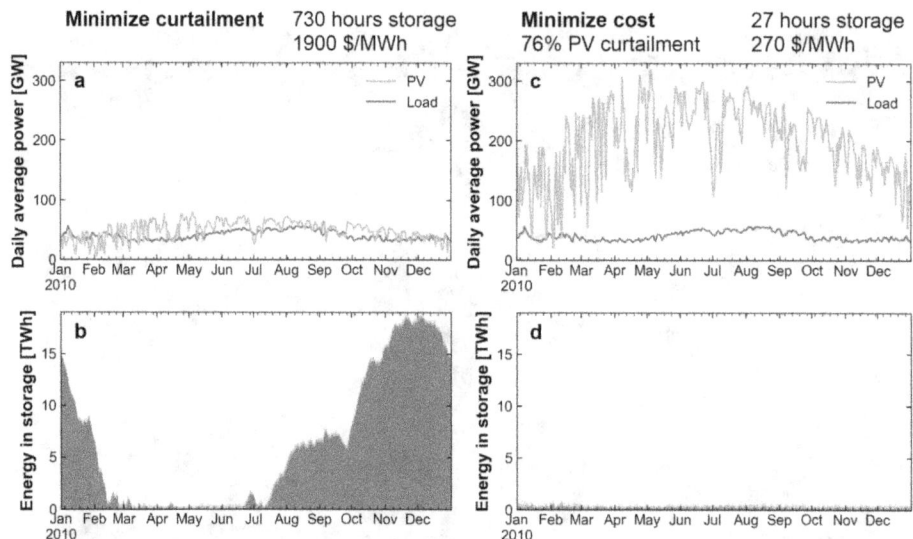

Figure 2.18. Meeting Texas electricity demand with PV and Li-ion batteries. (a) daily average PV generation and (b) hourly energy in storage when minimizing PV curtailment; (c) daily average PV generation and (d) hourly energy in storage when minimizing the combined cost of PV and storage. Storage duration is quantified in terms of hours of average system-wide demand; electricity cost is the annualized capacity cost of solar and storage divided by the cumulative demand over the simulation period. Hourly demand is from ERCOT for the years 2007–2013, here showing only 2010 (Electric Reliability Council of Texas (2020)); utility-scale PV and Li-ion battery costs are projections for 2030 from the NREL Annual Technology Baseline (National Renewable Energy Laboratory 2019).

that is, it is likely to be cheaper to size the system PV capacity to meet winter demand and curtail PV generation in the summer than to attempt to store excess PV generation from the summer for use in the winter (Perez *et al* 2019). Figure 2.18 illustrates this concept using a simplified model of the Texas system in 2030.

If we attempt to meet historical hourly demand in Texas using only PV and Li-ion batteries while minimizing PV curtailment, the necessary energy capacity of storage is equivalent to roughly 730 h of average system electricity demand. If we instead size the PV and storage capacity to minimize cost, the required energy capacity of storage drops to roughly 27 h of average system demand, and the average system cost of electricity drops by ~85%. Two important caveats are necessary for this admittedly simple example: first, it would be much cheaper to meet demand using a portfolio of zero-carbon generation technologies including wind, hydropower, and nuclear than to rely solely on PV; second, if cheap long-duration storage technologies such as hydrogen and pumped-hydro storage are available at large scale, the optimal solution would feature less overbuilding and curtailment of PV and more reliance on storage.

Advanced inverters: While the strategies described above address the challenges of PV intermittency at the ~1 h timescale, additional developments are necessary to ensure the transient stability of power systems with high penetrations of PV generation. Traditionally, power systems have been dominated by 'synchronous' generators employing large spinning turbines that are tightly coupled to each other through the synchronized AC power network. In these systems, if a single generator or power line trips offline, the 'inertia' provided by the rest of the generators on the system (the literal angular momentum of large, heavy rotating turbines) would pick up the slack in the critical few seconds between the incidence of the fault and the response of automatic generator controls, helping keep the system synchronized at the standard frequency (60 Hz in the North America and parts of South America and Asia; 50 Hz in most of the rest of the world). 'Asynchronous' generators such as PV, wind, and battery storage, which are coupled to the power system using solid-state inverters, do not provide this physical inertia.

Almost all inverters in use today are 'grid-following'—they count on the rest of the grid to provide a stable voltage and frequency while adjusting their own current and phase angle to match. A power system based entirely on grid-following inverters cannot achieve stable operation. Achieving a zero-carbon power system will thus require one of two approaches. In systems with large amounts of geothermal, hydropower, or nuclear power, grid-following inverters for PV, wind, and batteries could continue to be used, perhaps with a greater reliance on 'synchronous condensers' (large electrical motors that spin in sync with the grid, providing spinning inertia without generating power themselves) to maintain stability during times of high (but less than 100%) asynchronous penetration. Systems that reach 100% penetration of asynchronous generators would instead require 'grid-forming' inverters, which are designed to generate their own stable voltage and frequency, effectively simulating the performance of synchronous generators (Ellis 2018, Hoke et al 2018, National Renewable Energy Laboratory 2014).

While all of these approaches can increase the competitiveness of PV, the task of decarbonization will always become easier as the cost of primary energy production is reduced. We therefore turn now to the task of the PV scientist or engineer—decreasing the LCOE of PV.

References

Baker E, Fowlie M, Lemoine D and Reynolds S S 2013 The economics of solar electricity *Annu. Rev. Resour. Econ.* **5** 387–426

Bird L, Milligan M and Lew D 2013 Integrating variable renewable energy: challenges and solutions https://doi.org/10.2172/1097911

Bolinger M, Seel J and LaCommare K H 2017 Utility-scale solar 2016: an empirical analysis of project cost, performance, and pricing trends in the United States (Report number LBNL-2001055) Lawrence Berkeley National Laboratory (LBNL)

Borenstein S 2012 The private and public economics of renewable electricity generation *J. Econ. Perspect.* **26** 67–92

Brown P R and O'Sullivan F M 2019 Shaping photovoltaic array output to align with changing wholesale electricity price profiles *Appl. Energy* **256** 113734

CAISO 2017 California ISO Open Access Same-time Information System (OASIS)

California ISO—Renewables and emissions reports [www document] 2017 URL http://caiso.com/market/Pages/ReportsBulletins/RenewablesReporting.aspx (accessed 7.3.19)

Delucchi M A and Jacobson M Z 2011 Providing all global energy with wind, water, and solar power, Part II: reliability, system and transmission costs, and policies *Energy Policy* **39** 1170–90

Denholm P and Margolis R M 2007a Evaluating the limits of solar photovoltaics (PV) in traditional electric power systems *Energy Policy* **35** 2852–61

Denholm P and Margolis R M 2007b Evaluating the limits of solar photovoltaics (PV) in electric power systems utilizing energy storage and other enabling technologies *Energy Policy* **35** 4424–33

EIA 2012 Electric generator dispatch depends on system demand and the relative cost of operation *Today in Energy* https://www.eia.gov/todayinenergy/detail.php?id=7590

EIA 2017a Levelized cost and levelized avoided cost of new generation resources in the annual energy outlook 2017

EIA 2017b Rising solar generation in California coincides with negative wholesale electricity prices *Today Energy* https://www.eia.gov/todayinenergy/detail.php?id=30692

Electric Reliability Council of Texas 2020 Load http://www.ercot.org/gridinfo/load/

Ellis A 2018 Grid forming inverters in interconnected systems (SAND2018-11666C) Sandia National Laboratories https://der-lab.net/wp-content/uploads/2018/11/Ellis_GFI-Vienna.pdf

Fu R, Feldman D J, Margolis R M, Woodhouse M A and Ardani K B 2017 U.S. Solar Photovoltaic System Cost Benchmark: Q1 2017 (Report number NREL/TP-6A20-68925) National Renewable Energy Laboratory (NREL)

Heide D, Greiner M, von Bremen L and Hoffmann C 2011 Reduced storage and balancing needs in a fully renewable European power system with excess wind and solar power generation *Ren. Energy* **36** 2515–23

Hirth L 2013 The market value of variable renewables: the effect of solar wind power variability on their relative price *Energy Econ.* **38** 218–36

Hoke A, Giraldez J, Palmintier B, Ifuku E, Asano M, Ueda R and Symko-Davies M 2018 Setting the smart solar standard: collaborations between Hawaiian electric and the National Renewable Energy Laboratory *IEEE Power Energ. Mag.* **16** 18–29

International Technology Roadmap for Photovoltaic, 2012 Results 2013 ITRPV

International Technology Roadmap for Photovoltaic, 2018 Results 2019 ITRPV

Jean J, Brown P R, Jaffe R L, Buonassisi T and Bulović V 2015 Pathways for solar photovoltaics *Energy Environ. Sci.* **8** 1200–19

Jenkins J D, Zhou Z, Ponciroli R, Vilim R B, Ganda F, de Sisternes F and Botterud A 2018 The benefits of nuclear flexibility in power system operations with renewable energy *Appl. Energy* **222** 872–84

Joskow P L 2011 Comparing the costs of intermittent and dispatchable electricity generating technologies *Am. Econ. Rev.* **101** 238–41

Kempton W, Pimenta F M, Veron D E and Colle B A 2010 Electric power from offshore wind via synoptic-scale interconnection *Proc. Natl Acad. Sci. USA* **107** 7240–5

Kreifels N, Mayer J N, Burger B and Wittwer C 2014 Analysis of photovoltaics and wind power in future renewable energy scenarios *Energy Technol.* **2** 29–33

Lazard 2018 Lazard's Levelized Cost of Energy Analysis—Version 12.0

MacDonald A E, Clack C T M, Alexander A, Dunbar A, Wilczak J and Xie Y 2016 Future cost-competitive electricity systems and their impact on US CO_2 emissions *Nat. Clim. Chang.* **6** 526

Mills A and Wiser R 2010 Implications of wide-area geographic diversity for short-term variability of solar power (Report number LBNL-3884E) Lawrence Berkeley National Laboratory (LBNL) https://doi.org/10.2172/986925

Mills A and Wiser R 2012 Changes in the economic value of variable generation at high penetration levels: a pilot case study of california (Report number LBNL-5445E) Lawrence Berkeley National Laboratory (LBNL) https://doi.org/10.2172/1183176

Mills A D and Wiser R H 2015 Strategies to mitigate declines in the economic value of wind and solar at high penetration in California *Appl. Energy* **147** 269–78

Millstein D, Wiser R, Bolinger M and Barbose G 2017 The climate and air-quality benefits of wind and solar power in the United States *Nat. Energy* **2** 17134

National Renewable Energy Laboratory 2003 Reference Air Mass 1.5 Spectra 2003 https://www.nrel.gov/grid/solar-resource/spectra-am1.5.html (accessed 7.3.19)

National Renewable Energy Laboratory 2014 Advanced inverter functions to support high levels of distributed solar. NREL

National Renewable Energy Laboratory 2019 Annual Technology Baseline NREL https://atb.nrel.gov/electricity/2019/data.html

Needleman D B, Poindexter J R, Kurchin R C, Marius Peters I, Wilson G and Buonassisi T 2016 Economically sustainable scaling of photovoltaics to meet climate targets *Energy Environ. Sci.* **9** 2122–9

Perez M, Perez R, Rábago K R and Putnam M 2019 Overbuilding & curtailment: the cost-effective enablers of firm PV generation *Sol. Energy* **180** 412–22

Reichelstein S and Yorston M 2013 The prospects for cost competitive solar PV power *Energy Policy* **55** 117–27

Rhodes J D, King C, Gulen G, Olmstead S M, Dyer J S, Hebner R E, Beach F C, Edgar T F and Webber M E 2017 A geographically resolved method to estimate levelized power plant costs with environmental externalities *Energy Pol.* **102** 491–9

Schmidt O, Hawkes A, Gambhir A and Staffell I 2017 The future cost of electrical energy storage based on experience rates *Nat. Energy* **2** 17110

Sengupta M, Xie Y, Lopez A, Habte A, Maclaurin G and Shelby J 2018 The National Solar Radiation Data Base (NSRDB) *Ren. Sust. Energy Rev.* **89** 51–60

Sepulveda N A, Jenkins J D, de Sisternes F J and Lester R K 2018 The role of firm low-carbon electricity resources in deep decarbonization of power generation *Joule* **2** 2403–20

US Energy Information Administration 2013 Assessing the economic value of new utility-scale electricity generation projects

Woodhouse M A, Smith B, Ramdas A and Margolis R M 2019 Crystalline silicon photovoltaic module manufacturing costs and sustainable pricing: 1H 2018 benchmark and cost reduction road map (Report number NREL/TP-6A20-72134) National Renewable Energy Lab (NREL) https://www.osti.gov/biblio/1495719-crystalline-silicon-photovoltaic-module-manu-facturing-costs-sustainable-pricing-benchmark-cost-reduction-road-map

Chapter 3

Opportunities for emerging PV technologies

To achieve the lowest possible levelized cost of electricity (LCOE) in a given location, a PV system should have a low upfront cost ($ W^{-1}), long lifetime (years), high energy yield (kWh kW^{-1} per year), and low cost of capital (%) (Jones-Albertus *et al* 2016). These economic metrics directly drive the technology metrics and primary goals for PV research and development.

The **upfront system cost** is composed of module and non-module (balance-of-system, or BOS) costs. Many BOS cost components—including the cost of inverters, grid interconnection, inspection, permitting, taxes, and business overhead, among others—cannot be directly mitigated by improving PV technology. However, PV researchers can still help reduce both module and selected non-module costs. To reduce module costs, one can use low-cost materials, reduce raw materials usage (g W^{-1}), and reduce module manufacturing time, labor, equipment, and energy requirements ($ m^{-2} or $ W^{-1}). To reduce non-module costs, one can increase the module solar-to-electric power conversion efficiency (% or W m^{-2}) and simplify transportation and installation.

The **system lifetime** depends on the rate at which system performance degrades and informs projections of the system's practical operating lifetime. To extend the module and system lifetime—specified as T80, the time to reach 80% of the initial power output—PV technology research aims to reduce cell and module degradation rates (%/year). Performance stability must be evaluated under different environmental and operating conditions, such as elevated humidity and temperature, temperature and voltage cycling, and varying illumination levels.

The **energy yield** represents the real-world system performance, and is directly related to system performance metrics such as capacity factor and performance ratio (Jean *et al* 2015). To maximize energy yield, one can use solar tracking or develop strategies to improve cell and module power output under low light, off-angle illumination, spectral variations, high temperature, and other non-ideal conditions —ideally without increasing module costs substantially.

doi:10.1088/978-0-7503-2152-5ch3

The **cost of capital** represents the financial risk perceived by financiers of PV systems. A riskier investment corresponds to a higher discount rate in the LCOE analysis, as any electricity expected to be produced in the future is less certain and thus has a lower risk-adjusted value today. To help reduce financing costs, PV technologists can reduce uncertainty by proving cell and module performance under a wide range of operating conditions and improving predictions of actual system performance in the field.

These metrics can be traded off to achieve the same LCOE (figure 3.1). Increasing the efficiency or lifetime increases the allowed module price. Reducing module costs could allow new PV technologies to compete with c-Si on LCOE, even with a lower module efficiency. For example, a 0.40 $ W^{-1} c-Si module could reach the US Department of Energy's 2020 LCOE target with an efficiency of 20%. A 0.20 $ W^{-1} module with the same lifetime could reach the same LCOE with a lower efficiency of 16%.

Other metrics that are important for niche PV applications today include mechanical flexibility, weight-specific power or power-to-weight ratio (W kg^{-1}), and aesthetic characteristics such as color and visible transparency. While these metrics are not captured in a typical LCOE analysis, they could help reduce LCOE by enabling new modes of deployment—for example, rolling out lightweight and flexible solar panels on any surface or integrating transparent solar cells into insulated glass windows.

Table 3.1 summarizes how PV-related science and technology research can help reduce LCOE.

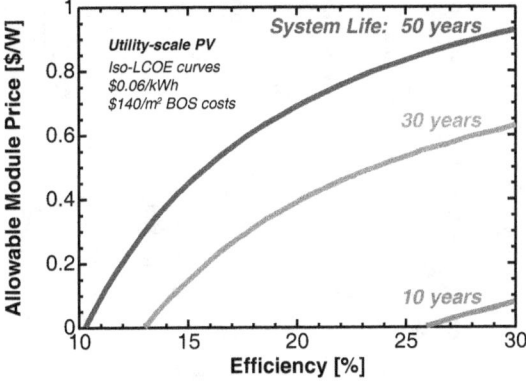

Figure 3.1. Iso-LCOE curves showing trade-off between PV performance metrics. The highest allowed module price is shown for utility-scale PV systems reaching the US Department of Energy's 2020 LCOE target of 0.06 $ kWh^{-1}, assuming a financial analysis period of 10, 30, and 50 years, a degradation rate of 2%, 0.2%, and 0.2% per year, and total system costs of 0.55, 1.10, and 1.40 $ W^{-1}, respectively. A nominal discount rate of 7% is assumed. BOS costs are fixed at 140 $ m^{-2}. Data are from DOE (Jones-Albertus *et al* 2016, Woodhouse *et al* 2016). Reproduced with permission from Jean (2017).

Table 3.1. Motivations for PV technology research with the long-term goal of reducing LCOE. PV researchers striving for real-world impact should view this table as a blueprint for answering the question, 'Why are you working on that problem?'.

Economic goal	Technology goal	Research goal	Example R&D project
Reduce module cost ($ W^{-1})	Reduce materials costs ($ m^{-2})	Develop lower-cost, higher-yield material synthesis and purification methods	New methods for automated QD synthesis and crashout (Pan *et al* 2013)
		Use lower-cost layers in the device stack	Metal oxide electron and hole transport layers for perovskite solar cells (Calió *et al* 2016, Chen *et al* 2015)
	Reduce raw materials usage (g W^{-1})	Use thinner absorber and transport layers	Thin inorganic charge extraction layers for perovskite solar cells (Chen *et al* 2015)
		Increase deposition yield	Close-space sublimation (CSS) of perovskite films (Guo *et al* 2016)
		Minimize solvent requirements	High-speed inkjet printing of QD and perovskite films (Li *et al* 2015)
		Develop new module architectures to reduce required active cell area	Engineering Stokes shift in QDs to reduce reabsorption loss in luminescent solar concentrators (LSCs) (Meinardi *et al* 2015)
	Reduce process time (increase throughput)	Increase deposition rate	Slot-die coating of perovskite solar cells (Hwang *et al* 2015)
		Demonstrate high-throughput substrate handling and deposition methods	Ambient roll-to-roll processing of flexible perovskite solar cells (Schmidt *et al* 2015)
		Increase module size (and power)	Techno-economic analysis of impact of module area on system cost and performance (Horowitz *et al* 2017)
	Reduce process labor requirements	Increase automation of cell and module processing steps	Fully automated roll-to-roll PV fabrication and testing systems (Krebs *et al* 2010)
	Reduce complexity of process equipment	Develop simple solution-based deposition techniques for absorber, transport layers, and electrodes	Blade coating of large-area perovskite films for high-efficiency solar cells (Deng *et al* 2016)

(Continued)

Table 3.1. (Continued)

Economic goal	Technology goal	Research goal	Example R&D project
	Reduce process energy requirements	Reduce process temperatures	Low-boiling-point solvents for rapid crystallization of perovskite films (Noel et al 2017)
		Eliminate process steps requiring high vacuum	Low-pressure chemical vapor deposition (CVD) of perovskite films (Luo et al 2015)
Reduce BOS cost ($ W^{-1})	Increase cell power conversion efficiency (%)	Develop new material compositions with improved optoelectronic properties (e.g. PL efficiency, mobility, carrier lifetime)	Multi-cation perovskite absorbers with narrow PL linewidths and high stabilized efficiencies (Saliba et al 2016)
		Understand and mitigate defect states contributing to non-radiative recombination	Passivation methods for reducing recombination at perovskite grain boundaries (Brenes et al 2017, deQuilettes et al 2015)
		Develop new device architectures to improve charge extraction and mitigate interface recombination	Ordered bulk heterojunctions with 1D nanostructures for improving charge extraction in QD solar cells (Jean et al 2013)
		Develop new PV technologies that exploit different physical mechanisms to surpass conventional efficiency limits	Multijunctions, upconversion, thermophotovoltaics, hot-carrier PV, and other 'third-generation' concepts (Green 2001)
	Increase module power conversion efficiency (%)	Improve large-area film uniformity and device performance	Monolithically integrated large-area perovskite solar cells (Matteocci et al 2014)
		Reduce front-surface reflection	Antireflection (AR) coatings on glass using stamped silica nanocylinder arrays (van de Groep et al 2015)
		Reduce series resistance losses	High-performance transparent conductors (Ellmer 2012)

	Develop reliable techniques for cell interconnection and module integration	Laser patterning for perovskite mini-modules with high geometric fill factor (Matteocci et al 2016, Razza et al 2016)
Simplify transportation and installation	Reduce substrate thickness	Thin, lightweight, flexible PV substrates for high specific power (Jean et al 2016)
	Develop PV technologies that can be integrated into existing structures	Neutral-color semitransparent perovskite solar cells for building-integrated PV (Eperon et al 2014)
Increase system lifetime (yr)	Reduce cell degradation rate (%/yr)	
	Improve intrinsic optoelectronic stability of each layer in the device stack under different environmental conditions	Compositional engineering of perovskite materials for reduced hysteresis (Jeon et al 2015)
	Understand mechanisms contributing to cell degradation and instability	Role of structure, environmental, and thermal stability in perovskite solar cell performance instability (Leijtens et al 2017)
	Quantify effect of environmental parameters on cell performance	Measuring the temperature coefficient of perovskite solar cells (Habisreutinger et al 2014)
	Develop more stable, self-encapsulating device architectures	Solution-processed metal-oxide transport layers for intimate encapsulation of perovskite absorbers (You et al 2015)
Reduce module degradation rate (%/yr)	Improve performance and reduce cost of moisture and oxygen barrier solutions	Flexible ultra-barrier coatings with low water vapor transmission rates (Burrows et al 2001)
	Develop improved encapsulation schemes for thin-film modules	Evaluation and modeling of edge-seal materials for PV applications (Kempe et al 2010)
	Understand mechanisms contributing to module degradation	Degradation mechanisms observed in deployed PV modules (Dhere 2005)

(Continued)

Table 3.1. (*Continued*)

Economic goal	Technology goal	Research goal	Example R&D project
Increase energy yield (kWh kW^{-1})	Increase module power output under real-world conditions	Understand effect of varying light intensity, spectrum, temperature, and humidity on PV power output	Modeling tandem perovskite PV energy yield in different locations and climates (Hörantner and Snaith 2017)
		Develop new module deployment strategies to increase power output per unit area	Solar tracking methods for improving PV energy output (Mousazadeh et al 2009)
Reduce financing cost (%)	Reduce uncertainty about module lifetime in the field	Validate module performance under different environmental conditions	Qualification of thin-film mini-modules based on stability test standards (Roesch et al 2015)
		Demonstrate long-term module reliability in deployed PV systems	Analysis of technology and climate trends in PV module degradation (Jordan and Kurtz 2013)
	Resolve policy and EHS issues associated with module use and disposal	Identify non-toxic alternatives to lead-based absorber materials	Bismuth-based double perovskite compounds for PV applications (Giustino and Snaith 2016)

3.1 The case for emerging PV technologies

Using the LCOE-related metrics above, today's crystalline silicon PV technology is hard to beat. After decades of development and government support, c-Si PV has reached record module efficiencies above 24% (Green *et al* 2017), production module efficiencies of 16%–20% (Barbose *et al* 2017, Philips and Warmuth 2019), field-proven lifetimes of over 25 years, module prices below 0.30 $ W^{-1} (International Technology Roadmap for Photovoltaic 2018), and utility-scale system prices below 1 $/W$_{dc}$ in many countries (Gallagher 2017). These metrics produce an unsubsidized LCOE of 0.12–0.16 $ kWh^{-1}, 0.09–0.12 $ kWh^{-1}, and 0.04–0.06 $ kWh^{-1} for residential, commercial, and utility-scale systems in the United States (Fu *et al* 2018) —already competitive with coal, gas, and other electric generation sources in many locations (figure 2.8).

It is no surprise that silicon dominates the PV market today, with a market share of over 90% (Philips and Warmuth 2019). Commercial thin films including cadmium telluride (CdTe), copper indium gallium diselenide (CIGS), and hydrogenated amorphous silicon (a-Si:H) make up the remainder. CdTe is the only thin-film technology with appreciable market share for grid-connected PV applications today. The rapid growth of the c-Si PV industry since 2010—driven by both worldwide consumer demand and Chinese support for domestic manufacturing—has produced massive economies of scale, continuous technological improvement, and a global supply chain. These factors add up to an incumbent cost and scaling advantage for c-Si PV that is difficult for other technologies to overcome. Given the urgency of climate change, we should deploy c-Si and existing commercial PV technologies far and wide today, even as we continue to develop new technologies.

But why work on emerging PV technologies at all, if silicon is so good already?

There are several unfavorable characteristics of today's PV industry and market landscape that each represent a market need and an opportunity for new technologies. Furthermore, looking beyond LCOE, one can envision emerging PV technologies enabling new applications with their advantages in weight, flexibility, and aesthetics.

First, the low profit margins and relatively high capital expenditures (capex) required for PV cell and module manufacturing make it difficult—and risky—for companies to rapidly and sustainably scale up manufacturing capacity (Powell *et al* 2015). This creates an opportunity for new technologies with significantly lower capex to enter the market and quickly capture market share. Low capex per peak watt can be achieved by simplifying manufacturing processes, reducing process temperatures, and increasing throughput. The high capex investment for c-Si PV manufacturing ($0.30 to $0.40 per watt of annual manufacturing capacity) means that a fully integrated c-Si PV factory with annual output of 2 GW would cost $600M or more to build (Woodhouse *et al* 2019), comparable to the quarterly revenue of the world's largest PV manufacturers. In contrast, flexible PV modules manufactured using high-throughput roll-to-roll processing could potentially reach a much lower capex—anecdotally well below $100M for a multi-gigawatt-scale production line.

Second, while c-Si PV modules have become extraordinarily cheap, non-module costs remain high, leading to higher overall system costs and LCOE (figure 2.7). A new PV technology with a higher efficiency or a more easily installed module format —for example, a lightweight, flexible, and rollable module—could potentially reduce BOS costs and provide a system-level cost advantage over c-Si. That said, the marginal cost savings decrease with increasing efficiency: at higher efficiencies, greater cost savings can typically be achieved by installing larger systems than by increasing module efficiency further (figure 3.2).

Commercial thin-film PV technologies such as CdTe and CIGS, which are already manufactured and deployed at large scale, can address some of the issues above. For example, CdTe modules on glass can reach low $ W^{-1} costs and are competitive with c-Si modules today. CIGS modules on stainless steel or specialized polymer foils can be efficient and flexible. For both technologies, however, materials scarcity is a major constraint to future large-scale deployment (figure 3.3). Tellurium and indium are scarce and produced at low volumes—as a result, CdTe and CIGS PV cannot account for more than 3% and 10%, respectively, of global electricity generation by 2030 without exceeding the maximum historically observed growth rates for metals production (Kavlak *et al* 2015). Furthermore, high process temperatures are needed to produce efficient CdTe and CIGS cells (figure 3.4). This constraint precludes the use of low-cost, lightweight plastic substrates and increases manufacturing capex.

Emerging thin-film PV technologies could reshape the solar landscape by exploiting all of the above opportunities. New PV absorber materials such as metal halide perovskites and colloidal quantum dots are structurally complex but relatively simple to process (figure 3.5). Importantly, these new materials appear

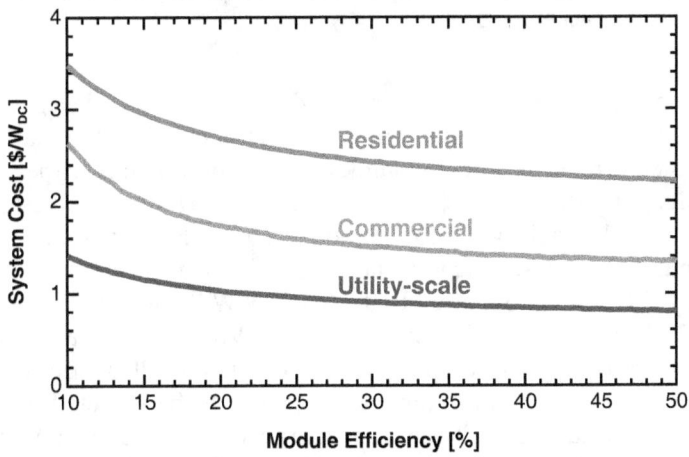

Figure 3.2. Impact of module efficiency on total system cost. Data are modeled values for US systems from NREL analysis (Fu *et al* 2017). Increasing module efficiency reduces area-dependent BOS and thus total system costs, especially at low efficiencies. At higher efficiencies, however, further efficiency improvements have a smaller impact on cost. Large utility-scale systems offer a much lower upfront cost and LCOE than residential and commercial systems.

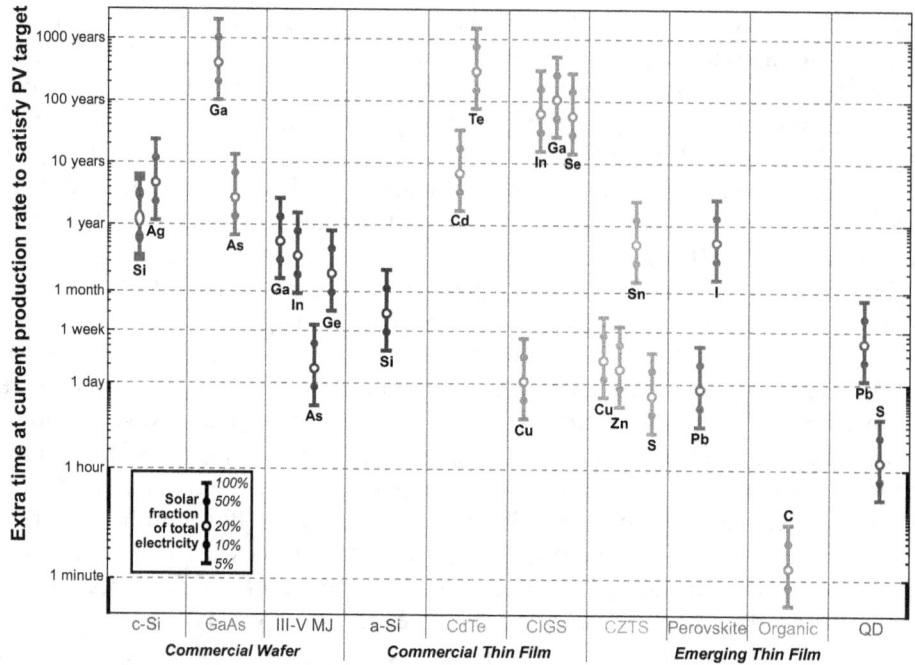

Figure 3.3. Critical material requirements for large-scale deployment of wafer-based, commercial thin-film, and emerging thin-film solar cell technologies. For each PV technology, we show the additional time at current global production rates required to produce enough key materials to satisfy 5%–100% of projected global electricity demand in 2050 solely with that technology, corresponding to a total installed capacity of 1.25–25 TW. Less additional time corresponds to less ramp-up in materials production required. Many emerging PV technologies can achieve terawatt-scale deployment without substantial growth in annual production of constituent elements. Boxes and oval for c-Si represent the range spanned by sc-Si and mc-Si technologies. Adapted with permission from Brown (2016).

to be more tolerant to impure precursors, defects, and high-level disorder than conventional semiconductors. Relaxing tolerances on the material purity and process parameters reduces process complexity, temperature, and capex. By employing high-volume-produced materials and high-throughput processes, emerging technologies could reach very low module costs. For example, unpublished manufacturing cost analyses suggest that perovskites can reach sustainable module prices below 0.20 \$ W^{-1}, compared to 0.32–0.35 \$ W^{-1} for today's c-Si technology and 0.20–0.25 \$ W^{-1} for future c-Si technology (Woodhouse *et al* 2019). We note that market prices for some classes of c-Si modules in 2020 are already in the 0.20–0.25 \$ W^{-1} range—this misalignment appears to be reflected in the negative (i.e. economically unsustainable) operating margins for most c-Si PV manufacturers, but may also be attributable to continued use of legacy or fully depreciated manufacturing equipment (Woodhouse *et al* 2019). In any case, while cost numbers tend to become outdated quickly in the fast-moving field of PV, the fundamental cost advantage of efficient thin-film modules seems to be persistent. Emerging

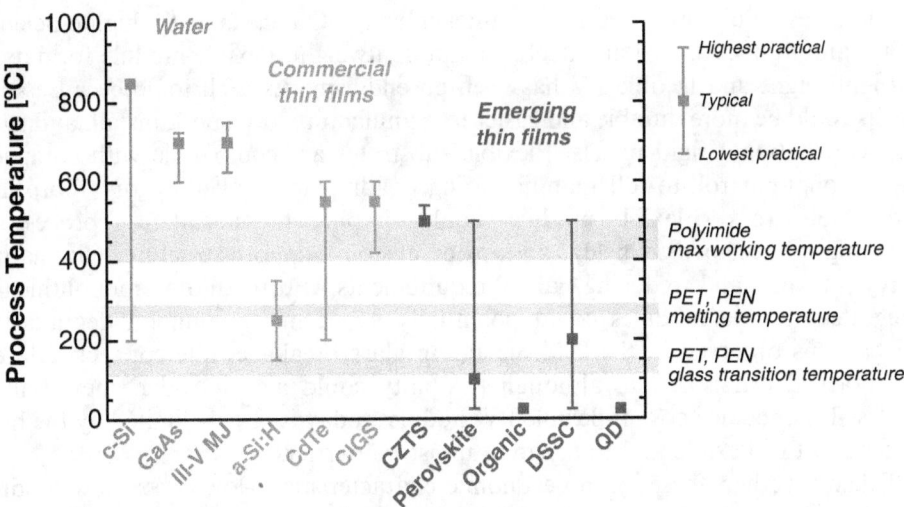

Figure 3.4. Typical process temperatures for PV technologies. Lower material purity requirements often lead to lower process temperatures. For example, c-Si PV typically uses elemental silicon refined to a purity of >99.9999% and process temperatures of over 800 °C. Commercial thin-film technologies such as CdTe and CIGS generally require precursor purities of >99.999% and process temperatures of 500 °C–600 °C to achieve high efficiencies. These high temperatures restrict flexible substrate choice to metal or polyimide foils. In contrast, emerging thin-film technologies commonly employ precursors with measurable purities on the order of 99% and process temperatures of 100 °C or lower. Low process temperatures for perovskite, organic, and QD solar cells enable the use of low-cost plastic substrates, such as polyethylene terephthalate and polyethylene naphthalate. Reproduced with permission from Jean (2017).

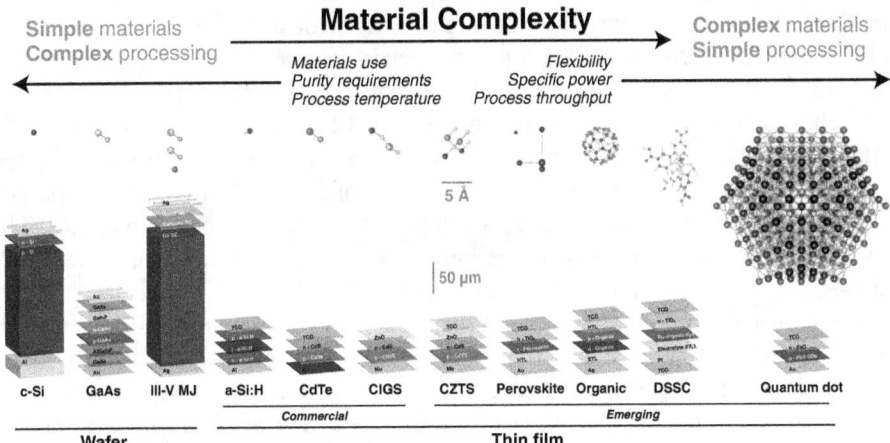

Figure 3.5. PV absorber material complexity. PV modules based on complex materials such as metal halide perovskites and colloidal QDs can be manufactured using simple, low-temperature, low-cost processes. Compared to commercial wafer and thin-film technologies, emerging thin-film PV technologies typically require less material and lower precursor purity and are compatible with lightweight, flexible substrates.

technologies could also reduce area-dependent BOS costs with high-efficiency (>30%) multijunction cell architectures and lightweight, flexible module formats.

Lightweight and flexible PV has been an enduring research topic because such panels could be more durable and easier to manufacture, transport, install, and stow than conventional rigid panels. Flexible substrates are compatible with compact, high-throughput roll-to-roll manufacturing, leading to smaller factory footprints and lower capex. Flexible modules could simplify the installation process to unrolling a sheet over a field, rooftop, or custom support structure, eliminating heavy racking and reducing labor requirements. Furthermore, monolithically integrated flexible modules could potentially avoid many common degradation mechanisms observed in the field, including glass breakage, interconnect failures, and cracked cell isolation, although flexibility could also introduce new failure modes. The modeled cost of flexible PV modules today is largely limited by the high present cost of flexible and transparent moisture barriers.

Taken together, these techno-economic characteristics—low capex, low module costs, and the potential to reduce BOS costs—may represent a substantial system cost and LCOE advantage for emerging PV technologies in the long term.

3.2 Entering the PV market

In the near term, emerging PV technologies cannot easily compete with c-Si on either an upfront module cost ($ W^{-1}) or LCOE ($ kWh^{-1}) basis. However, they may be able to compete by leveraging unique technological advantages such as high weight-specific power, flexibility, and visible transparency. These value propositions could enable new applications. For example, a high power-to-weight ratio is important for deployment on many flat commercial and industrial rooftops—which are not typically designed to support the additional weight of a PV array—as well as for rural deployment, aerospace applications, electric vehicles, telecommunications and Internet-of-things applications, and building-integrated PV (BIPV) (Reese *et al* 2018). The power-to-weight ratio is sensitive to the choice of substrate: With thin plastic substrates, emerging thin-film PV modules can reach much higher specific powers than conventional modules—on the order of 100–1000 W kg^{-1}, compared to 10 W kg^{-1} for conventional modules—even with a lower efficiency (figure 3.6). Thin, lightweight substrates are usually flexible as well, allowing additive or laminable solar power to be deployed on many surfaces (figure 3.7).

These high-value applications could serve as important stepping stones to large-scale manufacturing and deployment, providing the revenue needed to finance growth and the field experience needed to derisk the technology. But for researchers motivated by climate change mitigation, these novel PV applications are merely a means to an end. If the goal is to promote low-carbon electricity generation, new PV technologies must eventually reach a lower LCOE than incumbent technologies.

Toward that end, favorable economies of scale may be an important advantage of emerging PV technologies. Materials and process cost advantages could put emerging technologies on a new and lower experience (learning) curve, reducing the

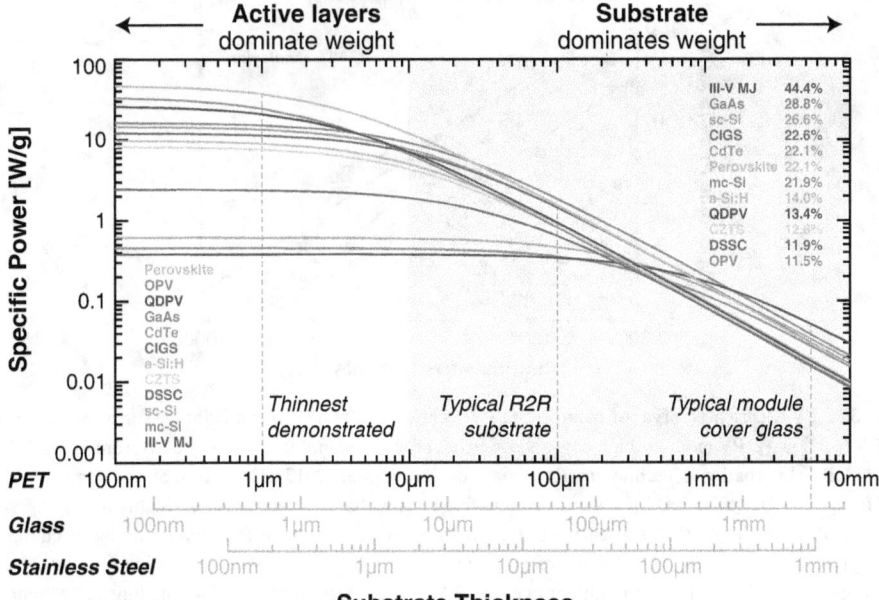

Figure 3.6. Specific power versus combined substrate and encapsulation thickness for different PV technologies. With thick substrates and encapsulation, the substrate dominates the total cell weight [g m^{-2}], and specific power is determined primarily by efficiency. With thin substrates and encapsulation, the cell dominates the total weight, and specific power is determined both by the device structure and the efficiency. Even with lower efficiencies, thin films can thus achieve much higher specific power than wafer technologies. Reproduced with permission from Jean (2017).

Figure 3.7. Reimagining rooftop solar PV systems using flexible, retractable modules. (a) Standard rigid PV module and rooftop system with discrete racking and wiring. (b) Flexible PV module with integrated wiring, enabling new modes of deployment.

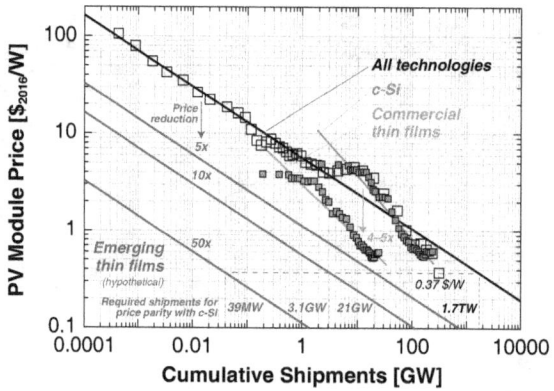

Figure 3.8. PV experience curve for commercial PV technologies and scaling advantage for hypothetical low-cost PV technology. PV module shipments and price data for all technologies (mostly c-Si) from 1976–2016 are from ITRPV (International Technology Roadmap for Photovoltaic 2017). Data for c-Si and commercial thin films from 2006 Q2 to 2015 Q4 are from Fraunhofer (Photovoltaics Report 2017). The historical experience curve for all technologies shows a learning rate (fractional price reduction for each doubling of cumulative shipment volume) of roughly 22%, while recent data for c-Si and thin films show learning rates of 29% and 25%, respectively. The 2016 c-Si module price of 0.37 $ W^{-1} is significantly below the long-term trend and corresponds to a cumulative deployment of 1.7 TW—much larger than the actual deployment of approximately 320 GW. Assuming the same learning rate as c-Si, a new thin-film technology with 5×, 10×, or 50× lower cost than c-Si at the same scale would need to reach 21 GW, 3.1 GW, or 39 MW of total shipments to reach a module price of 0.37 $ W^{-1}. Reduced process complexity may speed up learning for emerging technologies, reducing the shipment volume required to reach price parity with c-Si. Reproduced with permission from Jean (2017).

cumulative shipment volume required to reach price parity with c-Si modules (figure 3.8).

For example, a 5× price reduction at the same cumulative shipment volume would reduce the total production required to reach parity with 0.37 $ W^{-1} (as of 2016) c-Si modules by 81×, from 1.7 TW to 21 GW. A 10× or 50× price reduction would reduce the shipment volume for price parity to 3.1 GW (550×) or 39 MW (44 000×), respectively. As a point of reference, commercial thin films (CdTe, CIGS, and a-Si:H) have historically reached a price 4–5× lower than c-Si at a given cumulative shipment volume. Emerging thin-film modules could potentially be manufactured with even simpler, higher-throughput methods, enabling an even deeper price reduction. We should note that thin-film PV benefits from learning 'spillovers' from translatable c-Si production experience, making the choice of starting point for the thin-film experience curve somewhat ill-defined. Emerging PV technologies will similarly benefit from industry experience with c-Si and commercial thin-film technologies.

A lower shipment volume for price parity would allow emerging PV technologies to reach competitive scale faster—and with lower total capital investment—than earlier technologies. Even with lower costs at small scale, however, any new technology still has to move down the experience curve, with at least the first few gigawatts produced at a higher cost and sold at a higher price than commodity c-Si modules. To reach cost-competitiveness, new PV products will likely have to exploit the high-value attributes discussed above.

3.3 Accelerating market entry with module replacement

The path to grid-scale markets for any new PV technology is challenging: The expected future cost advantages are outweighed by high initial degradation rates, which reduces the modeled 25-year energy yield and system lifetime, and by the lack of outdoor performance data, which increases the cost of capital. As a result, new technologies are often deemed unbankable until decades of data have been accumulated. Thus the 25-year lifetime requirement for PV modules—itself a product of historical circumstance rather than technoeconomic need—contributes to technology lock-in, hindering the market entry of promising new PV technologies. Indeed, every single PV technology that has been deployed at large scale was invented—and made its first appearance on the NREL record efficiency chart—before 1980, over 40 years ago.

The conventional design life of a PV system is dictated by the module degradation rate, and installed modules are removed only upon acute failure or at the end of the system life. This operating strategy makes sense in the historical context of PV: Modules have traditionally comprised the majority of the system cost, and the dominant c-Si technology is engineered to operate reliably for decades (Jordan and Kurtz 2013). To reduce LCOE further, one could thus argue that we must extend system and module lifetimes to 30 or more years (Jones-Albertus *et al* 2016, Woodhouse *et al* 2016).

But the module lifetime need not dictate the system lifetime. For many PV systems today, modules constitute only a small fraction of the total cost (Fu *et al* 2018, Jean *et al* 2015). In this BOS-heavy cost structure—which despite substantial commercial effort is likely to persist for the foreseeable future—it may be that a 25-year module lifetime is not strictly necessary. Shorter-lived modules can be replaced one or more times during the system life at low cost, given that most of the BOS infrastructure is already in place. This is especially true when module performance and cost are improving rapidly, as in recent years.

In a 2019 study, we analyzed the impact of periodic module replacement on the LCOE of US PV systems (Jean *et al* 2019). Example system cost and performance data over a 30-year analysis period are shown in figure 3.9.

We found that a module replacement strategy allows a competitive levelized cost of electricity to be achieved with an initial module lifetime of less than 15 years, assuming backwards compatibility with the original system design (figure 3.10). The key condition is that the efficiency gain over installed modules—accounting for degradation—must be large enough to justify the added cost of replacement. Our analysis showed that new PV technologies may not need 25-year lifetimes to enter the market and achieve a competitive LCOE—assuming they have achieved a competitive module efficiency (e.g. $\geqslant 20\%$), cost (e.g. $\leqslant \$0.30/W$), and lifetime (e.g. $\geqslant 10$ years), and have the potential to improve further on all three metrics. This

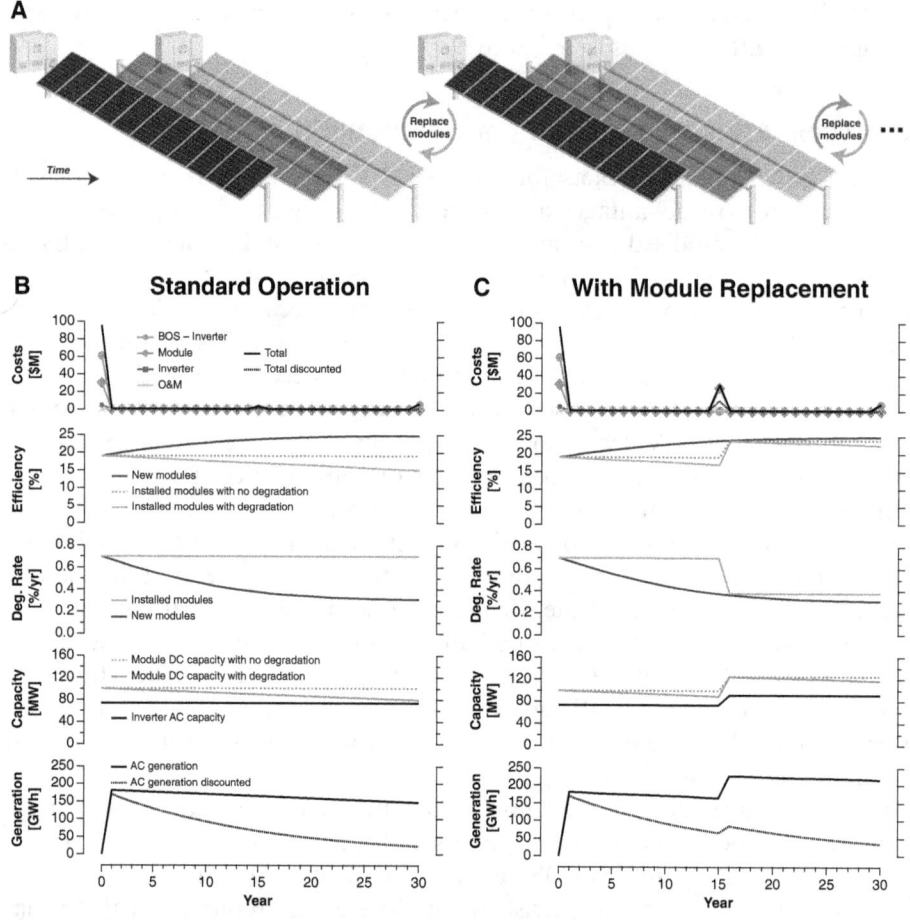

Figure 3.9. PV system operation with module replacement. (a) Schematic of periodic module replacement. (b) Time series data for a 100 MW utility-scale system employing a standard operating strategy with no module replacement. Year 0 is the time of initial installation. Module costs include module-related labor. Operation and maintenance (O&M) costs are incurred in every non-zero year. The PCE and degradation rate of newly manufactured modules improves toward limiting values of 25% and 0.3%/yr, respectively. Degradation reduces the installed PCE and DC capacity, leading to a linear decline in alternating-current (AC) output. Discounting of the AC generation is required for calculations but has no physical significance. (c) Time series data for the same system employing a module replacement strategy with a single replacement event at year 15. The installed DC capacity increases by 25% and degrades more slowly after module replacement, producing a higher AC output in years 16–30. The inverter is also upgraded to limit clipping losses. Reprinted from Jean *et al* (2019) with permission from Elsevier.

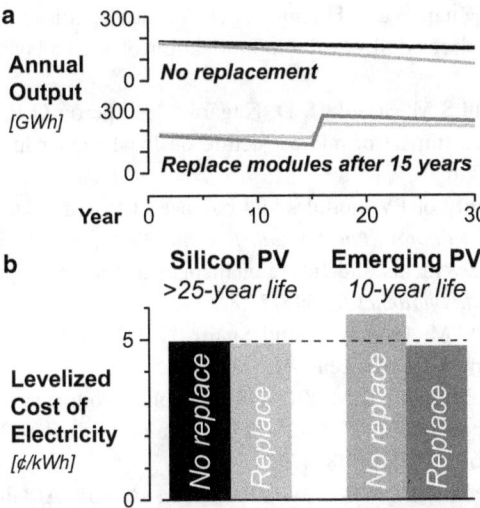

Figure 3.10. Module replacement to enable low LCOEs with short-lived modules. (a) Modeled annual AC output of 100 MW single-axis tracking utility-scale systems in Kansas City, Missouri, without module replacement and with replacement of all installed modules at year 15 of the assumed 30-year system life. The evolution of performance and cost parameters is projected for crystalline silicon modules (orange) and hypothetical high-efficiency, rapidly improving, but short-lived emerging PV modules (blue). At year 15, annual output steps up with the introduction of higher-efficiency modules. (b) Calculated LCOE for each system shown in (a). For crystalline silicon modules, the added cost of replacement nearly cancels out the increased output, leading to minimal change in LCOE. In contrast, for short-lived emerging PV modules, module replacement helps counteract inevitable degradation and leads to a substantial reduction in LCOE. Adapted from Jean *et al* (2019) with permission from Elsevier.

counterintuitive finding highlights a potential opportunity for the near-term market entry of emerging solar technologies that can reach extremely low costs but lack decades of field deployment experience.

References

Barbose G L, Darghouth N R, Millstein D, LaCommare K H, DiSanti N and Widiss R 2017 Tracking the Sun 10: the installed price of residential and non-residential photovoltaic systems in the United States. Lawrence Berkeley National Laboratory (LBNL)

Brenes R *et al* 2017 Metal halide perovskite polycrystalline films exhibiting properties of single crystals *Joule* **1** 155–67

Brown P R 2016 Energy level engineering in colloidal quantum dot solar cells *PhD Thesis* Massachusetts Institute of Technology

Burrows P E, Graff G L, Gross M E, Martin P M, Shi M K, Hall M, Mast E, Bonham C, Bennett W and Sullivan M B 2001 Ultra barrier flexible substrates for flat panel displays *Displays* **22** 65–9

Calió L, Kazim S, Grätzel M and Ahmad S 2016 Hole-transport materials for perovskite solar cells *Angew. Chem. Int. Ed.* **55** 14522–45

Chen W *et al* 2015 Efficient and stable large-area perovskite solar cells with inorganic charge extraction layers *Science* **350** 944–8

Deng Y, Dong Q, Bi C, Yuan Y and Huang J 2016 Air-stable, efficient mixed-cation perovskite solar cells with Cu electrode by scalable fabrication of active layer *Adv. Energy Mater.* **6** 1600372

deQuilettes D W, Vorpahl S M, Stranks S D, Nagaoka H, Eperon G E, Ziffer M E, Snaith H J and Ginger D S 2015 Impact of microstructure on local carrier lifetime in perovskite solar cells *Science* **348** 683–6

Dhere N G 2005 Reliability of PV modules and balance-of-system components *Conf. Record of the Thirty-First IEEE Photovoltaic Specialists Conf.* 2005 pp 1570–76

Ellmer K 2012 Past achievements and future challenges in the development of optically transparent electrodes *Nat. Photonics* **6** 809–17

Eperon G E, Burlakov V M, Goriely A and Snaith H J 2014 Neutral color semitransparent microstructured perovskite solar cells *ACS Nano* **8** 591–8

Fu R, Feldman D J and Margolis R M 2018 US solar photovoltaic system cost benchmark: Q1 2018 (Report number NREL/TP-6A20-72399) National Renewable Energy Laboratory (NREL) https://doi.org/10.2172/1483475

Fu R, Feldman D J, Margolis R M, Woodhouse M A and Ardani K B 2017 US solar photovoltaic system cost benchmark: Q1 2017 (Report number NREL/TP-6A20-68925) National Renewable Energy Laboratory (NREL) https://doi.org/10.2172/1390776

Gallagher B 2017 PV System Pricing H1 2017: Breakdowns and Forecasts [WWW Document]. GTM Research.

Giustino F and Snaith H J 2016 Toward lead-free perovskite solar cells *ACS Energy Lett.* **1** 1233–40

Green M A 2001 Third generation photovoltaics: ultra-high conversion efficiency at low cost *Prog. Photovoltaics Res. Appl.* **9** 123–35

Green M A, Hishikawa Y, Warta W, Dunlop E D, Levi D H, Hohl-Ebinger J and Ho-Baillie A W H 2017 Solar cell efficiency tables (version 50) *Prog. Photovoltaics Res. Appl.* **25** 668–76

Guo Q, Li C, Qiao W, Ma S, Wang F, Zhang B, Hu L, Dai S and Tan Z 2016 The growth of a $CH_3NH_3PbI_3$ thin film using simplified close space sublimation for efficient and large dimensional perovskite solar cells *Energy Environ. Sci.* **9** 1486–94

Habisreutinger S N, Leijtens T, Eperon G E, Stranks S D, Nicholas R J and Snaith H J 2014 Carbon nanotube/polymer composites as a highly stable hole collection layer in perovskite solar cells *Nano Lett.* **14** 5561–8

Hörantner M T and Snaith H J 2017 Predicting and optimising the energy yield of perovskite-on-silicon tandem solar cells under real world conditions *Energy Environ. Sci.* **10** 1983–93

Horowitz K A W, Fu R, Silverman T and Woodhouse M 2017 An analysis of the cost and performance of photovoltaic systems as a function of module area (Report number NREL/TP-6A20-67006) National Renewable Energy Laboratory (NREL)

Hwang K, Jung Y-S, Heo Y-J, Scholes F H, Watkins S E, Subbiah J, Jones D J, Kim D-Y and Vak D 2015 Toward large scale roll-to-roll production of fully printed perovskite solar cells *Adv. Mater.* **27** 1241–7

International Technology Roadmap for Photovoltaic, 2017 Results 2018 ITRPV

International Technology Roadmap for Photovoltaic, 2018 Results 2019 ITRPV

Jean J 2017 Performance limits for colloidal quantum dot and emerging thin-film solar cells *PhD Thesis* Massachusetts Institute of Technology http://hdl.handle.net/1721.1/111858

Jean J, Brown P R, Jaffe R L, Buonassisi T and Bulović V 2015 Pathways for solar photovoltaics *Energy Environ. Sci.* **8** 1200–19

Jean J, Chang S, Brown P R, Cheng J J, Rekemeyer P H, Bawendi M G, Gradečak S and Bulović V 2013 ZnO nanowire arrays for enhanced photocurrent in PbS quantum dot solar cells *Adv. Mater.* **25** 2790–6

Jean J, Wang A and Bulović V 2016 *In situ* vapor-deposited parylene substrates for ultra-thin, lightweight organic solar cells *Org. Electron.* **31** 120–6

Jean J, Woodhouse M and Bulović V 2019 Accelerating photovoltaic market entry with module replacement *Joule* **3** 2824–41

Jeon N J, Noh J H, Yang W S, Kim Y C, Ryu S, Seo J and Seok S I 2015 Compositional engineering of perovskite materials for high-performance solar cells *Nature* **517** 476–80

Jones-Albertus R, Feldman D, Fu R, Horowitz K and Woodhouse M 2016 Technology advances needed for photovoltaics to achieve widespread grid price parity *Prog. Photovoltaics Res. Appl.* **24** 1272–83

Jordan D C and Kurtz S R 2013 Photovoltaic degradation rates—an analytical review *Prog. Photovoltaics Res. Appl.* **21** 12–29

Kavlak G, McNerney J, Jaffe R L and Trancik J E 2015 Metal production requirements for rapid photovoltaics deployment *Energy Environ. Sci.* **8** 1651–59

Kempe M D, Dameron A A, Moricone T J and Reese M O 2010 Evaluation and modeling of edge-seal materials for photovoltaic applications *2010 35th IEEE Photovoltaic Specialists Conf.* pp 256–61

Krebs F C, Tromholt T and Jørgensen M 2010 Upscaling of polymer solar cell fabrication using full roll-to-roll processing *Nanoscale* **2** 873–86

Leijtens T, Bush K, Cheacharoen R, Beal R, Bowring A and McGehee M D 2017 Towards enabling stable lead halide perovskite solar cells: interplay between structural, environmental, and thermal stability *J. Mater. Chem.* A **5** 11483–500

Li S-G, Jiang K-J, Su M-J, Cui X-P, Huang J-H, Zhang Q-Q, Zhou X-Q, Yang L-M and Song Y-L 2015 Inkjet printing of $CH_3NH_3PbI_3$ on a mesoscopic TiO_2 film for highly efficient perovskite solar cells *J. Mater. Chem. A Mater. Energy Sustain* **3** 9092–7

Luo P, Liu Z, Xia W, Yuan C, Cheng J and Lu Y 2015 Uniform, stable, and efficient planar-heterojunction perovskite solar cells by facile low-pressure chemical vapor deposition under fully open-air conditions *ACS Appl. Mater. Interfaces* **7** 2708–14

Matteocci F *et al* 2016 High efficiency photovoltaic module based on mesoscopic organometal halide perovskite *Prog. Photovoltaics Res. Appl.* **24** 436–45

Matteocci F, Razza S, Di Giacomo F, Casaluci S, Mincuzzi G, Brown T M, D'Epifanio A, Licoccia S and Di Carlo A 2014 Solid-state solar modules based on mesoscopic organometal halide perovskite: a route towards the up-scaling process *Phys. Chem. Chem. Phys.* **16** 3918–23

Meinardi F, McDaniel H, Carulli F, Colombo A, Velizhanin K A, Makarov N S, Simonutti R, Klimov V I and Brovelli S 2015 Highly efficient large-area colourless luminescent solar concentrators using heavy-metal-free colloidal quantum dots *Nat. Nanotechnol.* **10** 878–85

Mousazadeh H, Keyhani A, Javadi A, Mobli H, Abrinia K and Sharifi A 2009 A review of principle and sun-tracking methods for maximizing solar systems output *Ren. Sust. Energy Rev.* **13** 1800–18

Noel N K, Habisreutinger S N, Wenger B, Klug M T, Hörantner M T, Johnston M B, Nicholas R J, Moore D T and Snaith H J 2017 A low viscosity, low boiling point, clean solvent system for the rapid crystallisation of highly specular perovskite films *Energy Environ. Sci.* **10** 145–52

Pan J, El-Ballouli A O, Rollny L, Voznyy O, Burlakov V M, Goriely A, Sargent E H and Bakr O M 2013 Automated synthesis of photovoltaic-quality colloidal quantum dots using separate nucleation and growth stages *ACS Nano* **7** 10158–66

Philips D S and Warmuth W 2019 Photovoltaics Report. Fraunhofer ISE

Photovoltaics Report 2017 Fraunhofer Institute for Solar Energy Systems

Powell D M, Fu R, Horowitz K, Basore P A, Woodhouse M and Buonassisi T 2015 The capital intensity of photovoltaics manufacturing: barrier to scale and opportunity for innovation *Energy Environ. Sci.* **8** 3395–408

Razza S, Castro-Hermosa S, Di Carlo A and Brown T M 2016 Research update: large-area deposition, coating, printing, and processing techniques for the upscaling of perovskite solar cell technology *APL Mater.* **4** 091508

Reese M O, Glynn S, Kempe M D, McGott D L, Dabney M S, Barnes T M, Booth S, Feldman D and Haegel N M 2018 Increasing markets and decreasing package weight for high-specific-power photovoltaics *Nat. Energy* **3** 1002–12

Roesch R, Faber T, von Hauff E, Brown T M, Lira-Cantu M and Hoppe H 2015 Procedures and practices for evaluating thin-film solar cell stability *Adv. Energy Mater.* 5

Saliba M *et al* 2016 Incorporation of rubidium cations into perovskite solar cells improves photovoltaic performance *Science* **354** 206–9

Schmidt T M, Larsen-Olsen T T, Carlé J E, Angmo D and Krebs F C 2015 Upscaling of perovskite solar cells: fully ambient roll processing of flexible perovskite solar cells with printed back electrodes *Adv. Energy Mater.* 5

van de Groep J, Spinelli P and Polman A 2015 Single-step soft-imprinted large-area nano-patterned antireflection coating *Nano Lett.* **15** 4223–8

Woodhouse M, Jones-Albertus R, Feldman D, Fu R, Horowitz K, Chung D, Jordan D and Kurtz S 2016 On the path to SunShot. The role of advancements in solar photovoltaic efficiency, reliability, and costs (Report number NREL/TP-6A20-65872). National Renewable Energy Laboratory (NREL). https://doi.org/10.2172/1253983

Woodhouse M A, Smith B, Ramdas A and Margolis R M 2019 Crystalline silicon photovoltaic module manufacturing costs and sustainable pricing: 1H 2018 benchmark and cost reduction road map (Report number NREL/TP-6A20-72134) National Renewable Energy Laboratory (NREL)

You J *et al* 2015 Improved air stability of perovskite solar cells via solution-processed metal oxide transport layers *Nat. Nanotechnol.* **11** 75–81

Chapter 4

Current research directions

Two emerging thin-film PV technologies—metal halide perovskites and colloidal quantum dots (QDs)—have developed rapidly in recent years thanks to intensive global research efforts (figure 4.1). We focus on these two technologies because of this rapid progress and their potential to achieve high efficiencies, which we discuss further in this chapter. Other emerging technologies such as organic photovoltaics (OPV) and dye-sensitized solar cells (DSSCs) are not discussed because their lower theoretical efficiency limits make grid-connected deployment less promising. That said, OPV and DSSC technologies can open up applications that are driven not by LCOE but by other metrics such as transparency, color, weight, and flexibility.

Perovskites are already nearing commercialization—with numerous active start-ups and established companies in the field—while QD solar cells require further efficiency improvements. In both fields, however, there remain important scientific questions and technological challenges.

Below we introduce perovskite and QD materials and their PV-related characteristics. We then describe the similarities and shared R&D challenges for perovskites and QD solar cells. We do not attempt to provide a comprehensive review of either field or to cite all relevant publications. Instead we give a researcher's perspective on these emerging technologies—what makes them scientifically interesting and technologically exciting, in light of the economic drivers discussed above.

The following discussion assumes that the reader has a basic understanding of semiconductor device physics, energy band diagrams, and solar cell operation. For supplemental reading, we recommend other resources on these topics (Honsberg and Bowden 1999, Luque and Hegedus 2011, Nelson 2003, Würfel and Würfel 2016).

4.1 Metal halide perovskites

The term 'perovskite' refers to any material that adopts the same crystal structure as calcium titanate—ABX_3, where A and B are cations and X is an anion (figure 4.2). The relative size of the ions determines whether the desired cubic perovskite

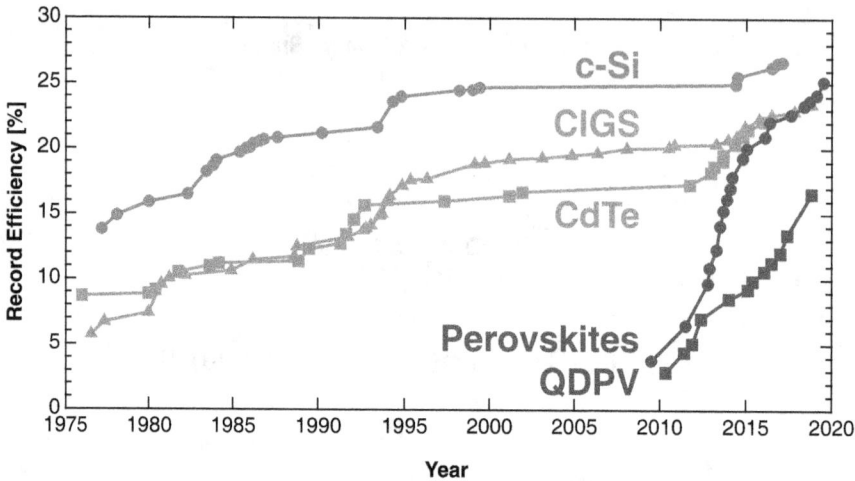

Figure 4.1. Record PV lab-cell efficiency trends. Commercial c-Si, CIGS, and CdTe solar cells reached efficiencies of over 20% after decades of development (National Renewable Energy Laboratory 2019). Emerging PV technologies such as metal halide perovskites and quantum dots have improved rapidly since 2010 (Jung and Park 2015), and perovskites have already surpassed CIGS and CdTe in efficiency.

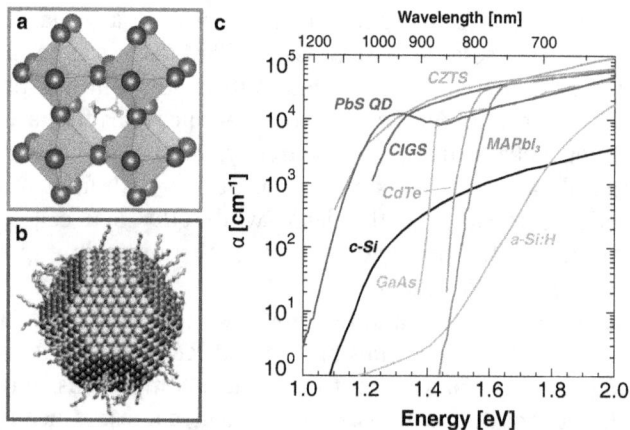

Figure 4.2. Crystal structures and absorption spectra of MAPbI$_3$ perovskite and PbS QD absorbers. (a) The perovskite ABX$_3$ unit cell consists of a methylammonium cation surrounded by iodide ions arranged in corner-sharing PbI$_6$ octahedra. (b) Lead sulfide QDs have diameters below 5 nm and are natively passivated with oleic acid and hydroxyl ligands. (c) Absorption spectra of commercial and emerging PV materials show the relatively sharp band edges of MAPbI$_3$ and PbS QD films. Panel (a) is licensed under CC BY 4.0 from NPG and Eames *et al* (2015). Panel (b) is reprinted from Thompson *et al* (2014) with permission from Springer Nature. Panel (c) is adapted with permission from Jean *et al* (2017). Copyright 2017 American Chemical Society.

structure can form[1]. The metal halide perovskites used in solar cells are ionic semiconductors typically synthesized by combining a metal salt BX_2 (e.g. lead(II) iodide, PbI_2) with an organic or inorganic halide salt AX (e.g. methylammonium iodide, MAI) in solution or vapor phase. There are several common compositional options for perovskites: methylammonium (MA, $CH_3NH_3^+$), formamidinium (FA, $(NH_2)_2CH^+$), and cesium (Cs^+) for the A-site cation; lead (Pb^{2+}) and tin (Sn^{2+}) for the B-site cation; and halides (I^-, Br^-, Cl^-) for the X-site anion. The composition most widely used in solar cells is the organic–inorganic perovskite $MAPbI_3$ (MAPI), which has a direct bandgap of roughly 1.6 eV.

The material properties of perovskites can be tuned easily by changing their composition. For example, the bandgap can be tuned from 1.6 eV down to 1.2 eV by incorporating tin (e.g. $FA_{0.6}MA_{0.4}Sn_{0.6}Pb_{0.4}I_3$ (Liao et al 2016) or $FA_{1-y}Cs_ySn_{1-x}Pb_xI_3$ (Eperon et al 2016)) or up to over 2.2 eV by incorporating bromine (e.g. $MAPbBr_3$ (Kojima et al 2009) or $FAPbBr_3$ (Eperon et al 2014)). Material and device stability can potentially be improved by using mixed compositions (e.g. $FA_{1-y}Cs_yPb(I_{1-x}Br_x)_3$) rather than pure $MAPbI_3$ (Leijtens et al 2017).

Organic–inorganic perovskite materials were first investigated in the early 1990s by Mitzi and colleagues but not used in solar cells (Mitzi 1999). In 2009, the first solar cells incorporating perovskites as light absorbers were demonstrated (Kojima et al 2009). In 2012, $MAPbI_3$ was shown by several groups to perform well as both the light absorber and charge transporter in standard PV architectures (Kim et al 2012, Lee et al 2012, Liu et al 2013). With this promising discovery—an inexpensive, solution-processed semiconductor with world-class optical and electronic properties —the race was on. Hundreds of research groups around the world jumped into the field of perovskite materials and devices, many with backgrounds in dye-sensitized, organic, polymer, and QD solar cells. Lessons learned from those fields helped perovskites advance faster than any previous PV technology. The history and current status of perovskite solar cell technology are thoroughly described in reviews by many leading researchers in the field (Ávila et al 2017, Berry et al 2017, Brenner et al 2016, Gratzel 2014, Green et al 2014, Huang et al 2017, Jung and Park 2015, Park et al 2016, Petrus et al 2017, Snaith 2013).

Metal halide perovskites possess many favorable properties for use in solar cells. Unlike conventional thin-film PV materials such as CdTe and CIGS, polycrystalline perovskite films can be formed at low temperatures (<150 °C), with many different deposition techniques yielding high power conversion efficiencies. Furthermore, the tunable direct bandgap allows sub-micron perovskite films to absorb nearly all above-bandgap light and enables tandem cell architectures with c-Si, CIGS, and other perovskites.

Most importantly, metal halide perovskites exhibit an optoelectronic quality— including efficient radiative recombination and charge transport—unprecedented for a low-temperature solution-processed material. Perovskites are by many metrics comparable to the best semiconductors available today (Huang et al 2017)—namely

---

[1] Some perovskites convert to tetragonal, orthorhombic, or other phases at lower temperatures.

epitaxially grown GaAs and other III–V materials. Unusual defect tolerance (Brandt *et al* 2015, Yin *et al* 2014) and low defect densities (Shi *et al* 2015) slow down non-radiative recombination, leading to long carrier lifetimes, high mobilities (of order $10 \text{ cm}^2 \text{ V}^{-1}\text{-s}^{-1}$ in thin films (Wehrenfennig *et al* 2014)), and long diffusion lengths (of order 1 μm (Stranks *et al* 2013)). Furthermore, sharp band edges—reflected in the low Urbach energy of 15 meV (De Wolf *et al* 2014)—contribute to a small Stokes shift (Tvingstedt *et al* 2014), high photoluminescence quantum efficiency (Deschler *et al* 2014), and low open-circuit voltage loss (Lee *et al* 2012), as well as a near-ideal ultimate PV efficiency limit (Jean *et al* 2017, Yablonovitch 2016). Photon recycling (Pazos-Outón *et al* 2016) and laser cooling (Ha *et al* 2015), possible only with a high-quality semiconductor, have already been achieved with perovskites.

Key fundamental limitations of today's leading perovskite materials include the use of toxic lead and unproven stability, which we discuss further below. Despite rapid progress on device stability, substantial further work is needed to achieve competitive cell and module lifetimes for terrestrial PV applications. If not resolved, low operational stability may impede the commercialization and grid-scale deployment of perovskite solar cells.

4.2 Colloidal quantum dots

Colloidal QDs are solution-processed nanocrystals sufficiently small to exhibit quantum-confinement effects—typically 1–10 nm in diameter (figure 4.2). Individual QDs can be considered as macromolecules, with optical and electronic properties tunable through their size and surface chemistry. QDs are typically synthesized in a one-pot batch-based method with chemical ligands physically adsorbed or chemically bonded to the QD surface. The colloidal ink is then deposited using a variety of techniques to form a semiconducting film.

Lead chalcogenides—PbS and PbSe—are the most common QD materials used in solar cells. The hot-injection synthesis route pioneered for visible-bandgap CdX QDs (X = S, Se, Te) by Murray, Norris, and Bawendi in 1993 (Murray *et al* 2000, 1993) was extended to near-infrared PbS QDs by Hines and Scholes in 2003 (Hines and Scholes 2003). Hot injection allows fine control of the QD growth process and size distribution, which is critical for many optoelectronic applications.

Key optoelectronic properties of QD films can be tuned by changing the QD size and ligands. The bandgap can be tuned over a wide range by changing the growth time and thus the average QD size. The Bohr radius of bound electron–hole pairs (excitons) is 20 nm in PbS and 46 nm in PbSe (Wise 2000), while a typical QD diameter is well below 10 nm. The resulting strong confinement widens the effective bandgap from the bulk value. For example, starting from the bulk PbS bandgap of 0.41 eV, PbS QD bandgaps—generally specified as the lowest-energy excitonic absorption peak—can be tuned from 0.7 eV to over 2 eV (600–1700 nm) (Grinolds *et al* 2015, Hines and Scholes 2003, Jasieniak *et al* 2011). Furthermore, the positions of the conduction and valence band edges and the Fermi level of QD films can be tuned by using different ligands with varying dipole moments (Brown *et al* 2014). The length of the ligand can affect charge transport, with shorter ligands leading to reduced inter-QD spacing and higher mobilities (Guyot-Sionnest 2012, Liu *et al* 2010). Finally, the chemical nature of the ligand can modify the QD stoichiometry and mid-gap trap density, which can dramatically affect PV performance (Ip *et al* 2012, Kim *et al* 2013). The most efficient PbS QD solar cells today use halide (typically iodide) ligands for the primary absorber layer and EDT ligands for the hole transport and electron blocking layer (Chuang *et al* 2014, Liu *et al* 2017).

The first major optoelectronic application of colloidal QDs was in light-emitting devices (LEDs), including phosphors used for displays and lighting. PbS QDs were first demonstrated in solar cells in the mid-2000s by Sargent, Nozik, and colleagues (Johnston *et al* 2008b, Klem *et al* 2007, Luther *et al* 2008, Maria *et al* 2005, McDonald *et al* 2005). Since 2010, QDPV power conversion efficiencies and air stability have improved rapidly with the development of halide-based ligand treatments and new device architectures (Chuang *et al* 2014, Ip *et al* 2012, Lan *et al* 2016, Liu *et al* 2017, Luther *et al* 2010, Ning *et al* 2014, Pattantyus-Abraham *et al* 2010, Tang *et al* 2011). More recently, solar cells employing perovskite (e.g. $Cs_{1-x}FA_xPbI_3$) QDs—whose bandgaps are tuned primarily through composition rather than size—have also reached high efficiencies (Hao *et al* 2020, Sanehira *et al* 2017, Swarnkar *et al* 2016), surpassing the lead chalcogenide QD efficiency record in 2017. The historical development of QD photovoltaics is thoroughly discussed in review papers and edited collections (Carey *et al* 2015a, Chernomordik *et al* 2017, Klimov 2017, Konstantatos and Sargent 2013, Kramer and Sargent 2014, Lan *et al* 2014, Sargent 2012, Tang and Sargent 2011, Yuan *et al* 2016).

Lead chalcogenide QDs possess several favorable properties for solar PV applications. They have a tunable direct bandgap, similar to perovskites. Unlike perovskites, however, their bandgaps can be tuned easily to 1 eV or lower, allowing tandem configurations with other PV technologies (Ip *et al* 2015). PbS QDs exhibit sharp band edges—with an Urbach energy of 22 meV for leading materials—yielding a theoretical PV efficiency limit above 30%, although the broad excitonic absorption peak may limit practical performance (Jean *et al* 2017). Most uniquely, the building blocks of a QD absorber (i.e. individual nanocrystals) are fully formed in solution prior to film deposition. QD pre-formation allows high-quality semiconducting films to be deposited from QD inks at room temperature, as no further crystallization is required.

Several key disadvantages limit the performance of QD solar cells today. Hopping charge transport leads to low charge mobilities in PbS QD films (of order 10^{-4} to 10^{-1} cm^2 V^{-1}·s^{-1} depending on the ligand (Jeong *et al* 2012, Johnston *et al* 2008a, Liu *et al* 2010, Osedach *et al* 2012, Tang *et al* 2011)), as well as short charge carrier diffusion lengths (of order 100 nm (Carey *et al* 2015b, Johnston *et al* 2008a)). Poor charge transport translates to inefficient charge extraction in PV devices. The high mid-gap trap densities (Bozyigit *et al* 2013, Jin *et al* 2016, Nelson and Zhu 2012, Zhang *et al* 2015) and strong electron–phonon interactions (Bozyigit *et al* 2016) make it difficult to mitigate non-radiative recombination, producing large V_{OC} deficits (Chuang *et al* 2015). On a practical note, cost modeling suggests that synthesizing colloidal QDs with existing methods is prohibitively expensive for PV applications (Jean *et al* 2018). It remains to be seen whether these limitations can be surmounted or if they will preclude commercialization of QD photovoltaics.

4.3 Perovskite and QD solar cells

The leading device architectures for perovskite and colloidal QD solar cells are common to the two technologies (figure 4.3). Both perovskites and QDs were initially used in a dye-sensitized solar cell (DSSC) architecture, with the perovskites or QDs adsorbed onto a mesoporous electron-transporting scaffold and infiltrated with a hole-transporting material. Only later were they shown to transport charge well, allowing the use of a conventional planar heterojunction architecture—similar

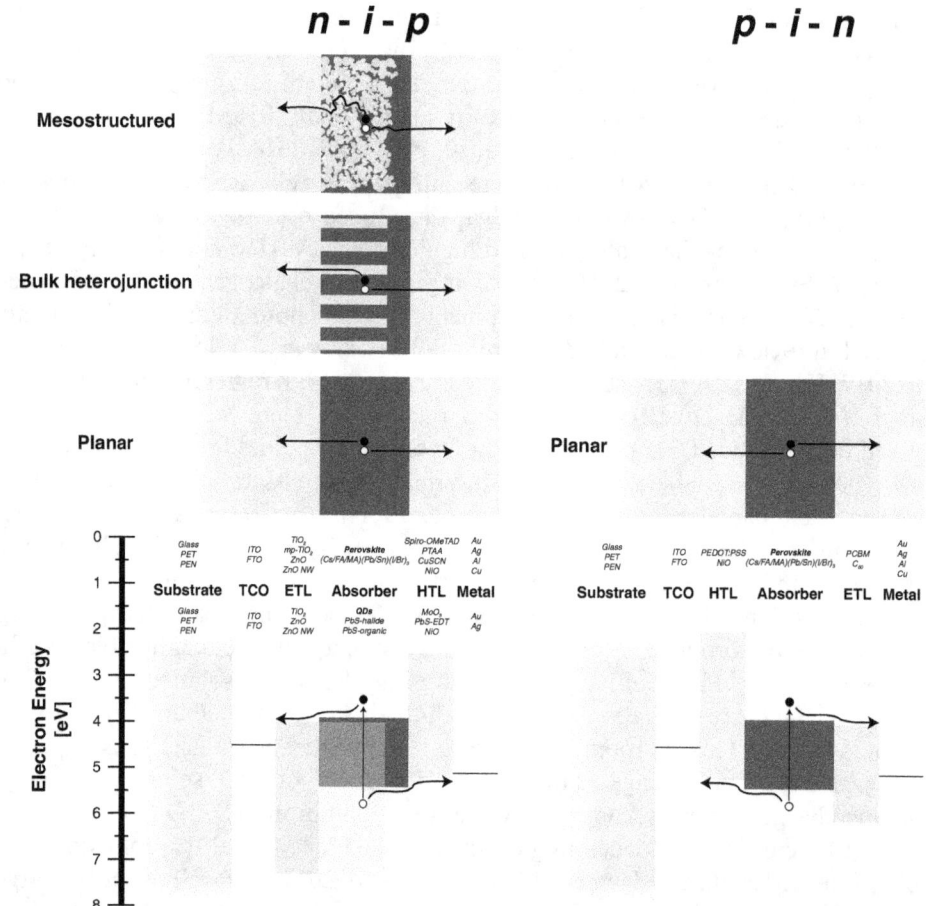

Figure 4.3. Common device structures and energy band diagrams for perovskite and QD solar cells. Both perovskites and QDs are typically used in a mesostructured, bulk heterojunction, or planar device architecture in a superstrate configuration, where incident light enters the device through a transparent substrate. The mesostructured and bulk heterojunction architectures use structured transport layers to assist conformal film coverage or to extract charge more efficiently—for example, to overcome short electron diffusion lengths in PbS QD films. Bulk heterojunctions can be ordered (as shown) or disordered (common in conjugated polymer and small-molecule organic solar cells). Common transparent conducting oxide (TCO), electron transport layer (ETL), absorber, hole transport layer (HTL), and metal electrode materials are listed. The flat-band energy levels shown in the band diagrams are representative values for each layer in the device stack.

to commercial thin-film PV technologies (CdTe and CIGS). However, some of the most efficient perovskite devices still employ a mesoporous TiO_2 scaffold to assist charge extraction. Many of the same transport layer and electrode materials have been investigated for perovskite and QD solar cells.

Perovskite and QD solar cells share several characteristics that could contribute to a low LCOE (table 4.1). Most importantly, both technology classes employ abundant, low-cost raw materials, are compatible with low-temperature, low-capex,

Table 4.1. Long-term potential of perovskites and QDs, with key LCOE metrics projected relative to crystalline silicon.

LCOE metric		Perovskite solar cells	QD solar cells
Upfront cost	Materials costs	Lower than c-Si	Uncertain
	Raw material usage	Lower than c-Si	Lower than c-Si
	Process time	Lower than c-Si	Lower than c-Si
	Labor requirements	Lower than c-Si	Lower than c-Si
	Process complexity	Lower than c-Si	Lower than c-Si
	Energy requirements	Lower than c-Si	Lower than c-Si
	Module efficiency	Higher than c-Si	Lower than c-Si
	Installation complexity	Same or lower than c-Si	Same or lower than c-Si
System lifetime	Degradation rate	Higher than c-Si	Higher than c-Si
Energy yield	Real-world power output	Uncertain	Uncertain
Cost of capital	Performance uncertainty	Higher than c-Si	Higher than c-Si
	Policy and EHS issues	Higher than c-Si	Higher than c-Si

high-throughput deposition techniques, and can potentially reach high efficiencies in a single-junction or tandem architecture.

The raw materials used in perovskite and QDs—carbon, iodine, bromine, lead, sulfur—are abundant in the Earth's crust and extracted in high volumes already (Jean *et al* 2015). High elemental abundance in a readily accessible form ensures that absolute material availability will not constrain terawatt-scale PV manufacturing and deployment. High production volumes reduce costs and help ensure supply. Furthermore, for perovskites, the chemical precursors are cheap and constitute only a small fraction of the total module cost, although we note that the substrate, encapsulation, and junction box remain significant cost components (Chang *et al* 2017, Song *et al* 2017).

Perovskite and QD films can be deposited at low temperature using high-throughput techniques. At the lab scale, the most common deposition technique used for both technologies is spincoating, which is not suitable for large-area films. Colloidal QDs naturally require solution methods, while perovskites are compatible with both solution and vapor methods. A variety of scalable deposition techniques have been explored, mostly for perovskites. Solution methods include blade coating, slot-die coating, microgravure printing, spray coating, and inkjet printing. Vapor deposition methods include thermal evaporation, close-space sublimation (CSS), and various chemical vapor deposition (CVD) methods.

Perovskite and QD absorbers have the potential to reach high PV power conversion efficiencies. Their low exciton binding energies—5–25 meV for perovskites at room temperature (Galkowski *et al* 2016, Miyata *et al* 2015) and <100 meV for PbS QDs (Choi *et al* 2010, Kang and Wise 1997)—allows spontaneous dissociation of photogenerated excitons into free carriers with minimal voltage loss (D'Innocenzo *et al* 2014, Talgorn *et al* 2011). In this regard, perovskites and QDs behave more similarly to conventional inorganic semiconductors than excitonic materials. Sharp band edges lead to high theoretical efficiency limits (Jean *et al* 2017). Furthermore, bandgap tunability—via material composition for perovskites and the quantum size effect for QDs—allows multijunction devices to be made with relative ease with either material class. Since multijunction cells are the only proven strategy for exceeding the Shockley–Queisser efficiency limit, the ability to exploit a multijunction architecture greatly improves the prospects of any new PV technology.

4.4 Tandem solar cells

One promising strategy for near-term commercialization is to combine a wide-bandgap perovskite cell with a low-bandgap perovskite, c-Si, or CIGS cell in a tandem architecture, which can achieve practical efficiencies over 30% and theoretical limits over 40% (figure 4.4) (Bailie *et al* 2015, Bailie and McGehee 2015, Berry *et al* 2017, Bush *et al* 2017, Eperon *et al* 2016, Hörantner *et al* 2017, Leijtens *et al* 2018, Todorov *et al* 2016). A multijunction cell is preferred over a single junction if its LCOE is lower than either the top or bottom cell operating independently (Peters *et al* 2016).

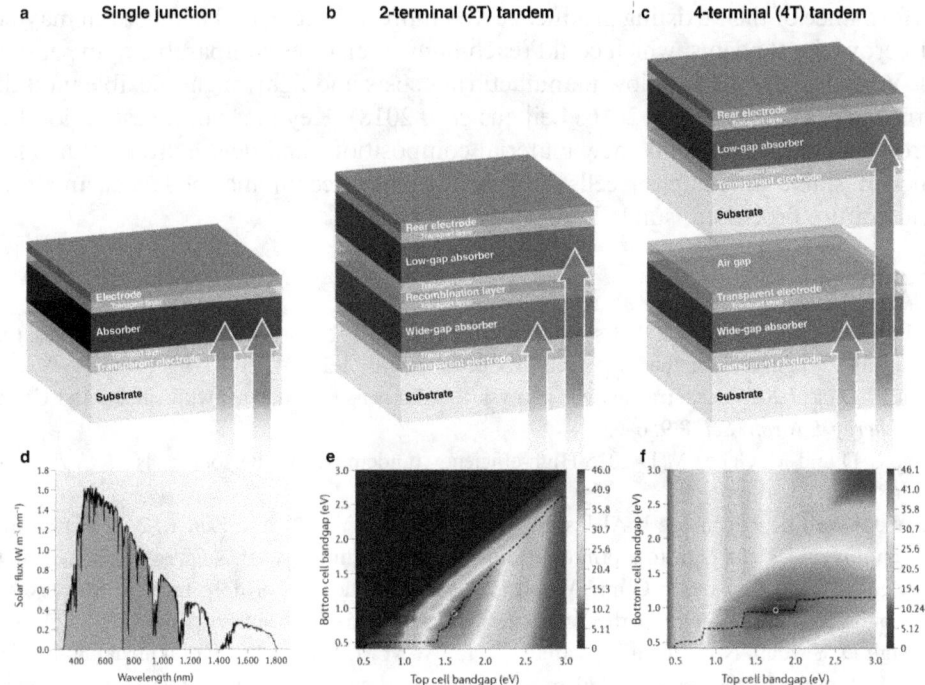

Figure 4.4. Tandem thin-film solar cell device structures and efficiency limits. Typical device structures are shown for (a) single-junction, (b) 2-terminal (2T) tandem, and (c) 4-terminal (4T) tandem solar cells. In a single-junction cell, all photons above the bandgap are absorbed in the single absorber layer. In a tandem cell, higher-energy (shorter-wavelength) photons are absorbed in the front wide-gap absorber, while lower-energy (longer-wavelength) photons pass through and are absorbed in the rear low-gap absorber. (d) The solar irradiance spectrum shows the spectral regions over which the two semiconductors can absorb. The theoretical maximum efficiency is shown as a function of front (top) and rear subcell bandgap for (e) 2T and (f) 4T tandem architectures, assuming no absorption losses. The dashed lines trace the peak efficiency for each front-cell bandgap, and the white circles indicate the maximum efficiencies. Panels (d)–(f) are reprinted from Eperon *et al* (2017) with permission. Copyright 2017 Springer Nature.

Tandem cells come in two primary configurations: 2-terminal (2T) and 4-terminal (4T). 2T or monolithic tandems consist of two cells of different bandgaps fabricated sequentially and connected in series, which requires that the subcell currents be matched. 4T tandems typically consist of two cells fabricated separately and mechanically stacked. Since 4T tandems can be connected in series or in parallel, they do not require current matching, which makes their theoretical efficiency limits less sensitive to the specific subcell bandgaps and opens up a wider design space (figure 4.4). While 4T tandems are easier to prototype, however, they employ 2 substrates and 3 transparent electrode layers (vs. 1 and 1–2 for 2T tandems), increasing parasitic absorption and reflection losses and likely leading to lower practical efficiencies.

The first large-scale commercial application of perovskites may come from c-Si PV manufacturers adopting a perovskite-silicon tandem architecture to boost the performance of their existing products. Even more attractive in the long term may be all-perovskite tandems, which could reach high efficiencies comparable to expensive III–V multijunctions with low manufacturing costs and lightweight, flexible module form factors (Eperon *et al* 2016, Leijtens *et al* 2018). Key areas of investigation for perovskite tandems include new material compositions and device architectures for efficient, stable wide-gap top cells, high-performance recombination layers, and low-temperature-processed, stable low-gap bottom cells.

References

Ávila J, Momblona C, Boix P P, Sessolo M and Bolink H J 2017 Vapor-deposited perovskites: the route to high-performance solar cell production? *Joule* **1** 1–12

Bailie C D *et al* 2015 Semi-transparent perovskite solar cells for tandems with silicon and CIGS *Energy Environ. Sci.* **8** 956–63

Bailie C D and McGehee M D 2015 High-efficiency tandem perovskite solar cells *MRS Bull.* **40** 681–6

Berry J J, van de Lagemaat J, Al-Jassim M M, Kurtz S, Yan Y and Zhu K 2017 Perovskite photovoltaics: the path to a printable terawatt-scale technology *ACS Energy Lett.* **2** 2540–4

Bozyigit D, Volk S, Yarema O and Wood V 2013 Quantification of deep traps in nanocrystal solids, their electronic properties, and their influence on device behavior *Nano Lett.* **13** 5284–8

Bozyigit D, Yazdani N, Yarema M, Yarema O, Lin W M M, Volk S, Vuttivorakulchai K, Luisier M, Juranyi F and Wood V 2016 Soft surfaces of nanomaterials enable strong phonon interactions *Nature* **531** 618–22

Brandt R E, Stevanović V, Ginley D S and Buonassisi T 2015 Identifying defect-tolerant semiconductors with high minority-carrier lifetimes: beyond hybrid lead halide perovskites *MRS Commun.* **5** 265–75

Brenner T M, Egger D A, Kronik L, Hodes G and Cahen D 2016 Hybrid organic—inorganic perovskites: low-cost semiconductors with intriguing charge-transport properties *Nat. Rev. Mater.* **1** 15007

Brown P R, Kim D, Lunt R R, Zhao N, Bawendi M G, Grossman J C and Bulović V 2014 Energy level modification in lead sulfide quantum dot thin films through ligand exchange *ACS Nano* **8** 5863–72

Bush K A *et al* 2017 23.6%-efficient monolithic perovskite/silicon tandem solar cells with improved stability *Nat. Energy* **2** 17009

Carey G H, Abdelhady A L, Ning Z, Thon S M, Bakr O M and Sargent E H 2015a Colloidal quantum dot solar cells *Chem. Rev.* **115** 12732–63

Carey G H, Levina L, Comin R, Voznyy O and Sargent E H 2015b Record charge carrier diffusion length in colloidal quantum dot solids via mutual dot-to-dot surface passivation *Adv. Mater.* **27** 3325–30

Chang N L, Yi Ho-Baillie A W, Basore P A, Young T L, Evans R and Egan R J 2017 A manufacturing cost estimation method with uncertainty analysis and its application to perovskite on glass photovoltaic modules *Prog. Photovoltaics Res. Appl.* **25** 390–405

Chernomordik B D, Marshall A R, Pach G F, Luther J M and Beard M C 2017 Quantum dot solar cell fabrication protocols *Chem. Mater.* **29** 189–98

Choi J J, Luria J, Hyun B-R, Bartnik A C, Sun L, Lim Y-F, Marohn J A, Wise F W and Hanrath T 2010 Photogenerated exciton dissociation in highly coupled lead salt nanocrystal assemblies *Nano Lett.* **10** 1805–11

Chuang C-H M, Brown P R, Bulović V and Bawendi M G 2014 Improved performance and stability in quantum dot solar cells through band alignment engineering *Nat. Mater.* **13** 796–801

Chuang C-H M, Maurano A, Brandt R E, Hwang G W, Jean J, Buonassisi T, Bulović V and Bawendi M G 2015 Open-circuit voltage deficit, radiative sub-bandgap states, and prospects in quantum dot solar cells *Nano Lett.* **15** 3286–94

D'Innocenzo V, Grancini G, Alcocer M J P, Kandada A R S, Stranks S D, Lee M M, Lanzani G, Snaith H J and Petrozza A 2014 Excitons versus free charges in organo-lead tri-halide perovskites *Nat. Commun.* **5** 3586

De Wolf S, Holovsky J, Moon S-J, Löper P, Niesen B, Ledinsky M, Haug F-J, Yum J-H and Ballif C 2014 Organometallic halide perovskites: sharp optical absorption edge and its relation to photovoltaic performance *J. Phys. Chem. Lett.* **5** 1035–9

Deschler F *et al* 2014 High photoluminescence efficiency and optically pumped lasing in solution-processed mixed halide perovskite semiconductors *J. Phys. Chem. Lett.* **5** 1421–6

Eames C, Frost J M, Barnes P R F, O'Regan B C, Walsh A and Islam M S 2015 Ionic transport in hybrid lead iodide perovskite solar cells *Nat. Commun.* **6** 7497

Eperon G E, Burlakov V M, Goriely A and Snaith H J 2014 Neutral color semitransparent microstructured perovskite solar cells *ACS Nano* **8** 591–8

Eperon G E, Hörantner M T and Snaith H J 2017 Metal halide perovskite tandem and multiple-junction photovoltaics *Nat. Rev. Chem.* **1** 0095

Eperon G E *et al* 2016 Perovskite–perovskite tandem photovoltaics with optimized bandgaps *Science* **354** 861–5

Eperon G E, Stranks S D, Menelaou C, Johnston M B, Herz L M and Snaith H J 2014 Formamidinium lead trihalide: a broadly tunable perovskite for efficient planar hetero-junction solar cells *Energy Environ. Sci.* **7** 982–8

Galkowski K *et al* 2016 Determination of the exciton binding energy and effective masses for methylammonium and formamidinium lead tri-halide perovskite semiconductors *Energy Environ. Sci.* **9** 962–70

Gratzel M 2014 The light and shade of perovskite solar cells *Nat. Mater.* **13** 838–42

Green M A, Ho-Baillie A and Snaith H J 2014 The emergence of perovskite solar cells *Nat. Photonics* **8** 506–14

Grinolds D D W, Brown P R, Harris D K, Bulovic V and Bawendi M G 2015 Quantum-dot size and thin-film dielectric constant: precision measurement and disparity with simple models *Nano Lett.* **15** 21–6

Guyot-Sionnest P 2012 Electrical transport in colloidal quantum dot films *J. Phys. Chem. Lett.* **3** 1169–75

Ha S-T, Shen C, Zhang J and Xiong Q 2015 Laser cooling of organic–inorganic lead halide perovskites *Nat. Photonics* **10** 115–21

Hao M *et al* 2020 Ligand-assisted cation-exchange engineering for high-efficiency colloidal $Cs_{1-x}FA_xPbI_3$ quantum dot solar cells with reduced phase segregation *Nat. Energy* **5** 79–88

Hines M A and Scholes G D 2003 Colloidal PbS nanocrystals with size-tunable near-infrared emission: observation of post-synthesis self-narrowing of the particle size distribution *Adv. Mater.* **15** 1844–9

Honsberg C and Bowden S 2019 Photovoltaics Education Website https://www.pveducation.org/pvcdrom/welcome-to-pvcdrom/instructions

Hörantner M T, Leijtens T, Ziffer M E, Eperon G E, Christoforo M G, McGehee M D and Snaith H J 2017 The potential of multijunction perovskite solar cells *ACS Energy Lett.* **2** 2506–13

Huang J, Yuan Y, Shao Y and Yan Y 2017 Understanding the physical properties of hybrid perovskites for photovoltaic applications *Nat. Rev. Mater.* **2** 17042

Ip A H, Kiani A, Kramer I J, Voznyy O, Movahed H F, Levina L, Adachi M M, Hoogland S and Sargent E H 2015 Infrared colloidal quantum dot photovoltaics via coupling enhancement and agglomeration suppression *ACS Nano* **9** 8833–42

Ip A H *et al* 2012 Hybrid passivated colloidal quantum dot solids *Nat. Nanotechnol.* **7** 577–82

Jasieniak J, Califano M and Watkins S E 2011 Size-dependent valence and conduction band-edge energies of semiconductor nanocrystals *ACS Nano* **5** 5888–902

Jean J, Brown P R, Jaffe R L, Buonassisi T and Bulović V 2015 Pathways for solar photovoltaics *Energy Environ. Sci.* **8** 1200–19

Jean J, Mahony T S, Bozyigit D, Sponseller M, Holovský J, Bawendi M G and Bulović V 2017 Radiative efficiency limit with band tailing exceeds 30% for quantum dot solar cells *ACS Energy Lett.* **2** 2614–24

Jean J, Xiao J, Nick R, Moody N, Nasilowski M, Bawendi M and Bulović V 2018 Synthesis cost dictates the commercial viability of lead sulfide and perovskite quantum dot photovoltaics *Energy Environ. Sci.* **11** 2295–305

Jeong K S *et al* 2012 Enhanced mobility-lifetime products in PbS colloidal quantum dot photovoltaics *ACS Nano* **6** 89–99

Jin Z, Wang A, Zhou Q, Wang Y and Wang J 2016 Detecting trap states in planar PbS colloidal quantum dot solar cells *Sci. Rep.* **6** 37106

Johnston K W, Pattantyus-Abraham A G, Clifford J P, Myrskog S H, MacNeil D D, Levina L and Sargent E H 2008b Schottky-quantum dot photovoltaics for efficient infrared power conversion *Appl. Phys. Lett.* **92** 151115

Johnston K W, Pattantyus-Abraham A G, Clifford J P, Myrskog S H, Hoogland S, Shukla H, Klem E J D, Levina L and Sargent E H 2008a Efficient Schottky-quantum-dot photovoltaics: the roles of depletion, drift, and diffusion *Appl. Phys. Lett.* **92** 122111

Jung H S and Park N-G 2015 Perovskite solar cells: from materials to devices *Small* **11** 10–25

Kang I and Wise F W 1997 Electronic structure and optical properties of PbS and PbSe quantum dots *J. Opt. Soc. Am. B, JOSAB* **14** 1632–46

Kim D, Kim D-H, Lee J-H and Grossman J C 2013 Impact of stoichiometry on the electronic structure of PbS quantum dots *Phys. Rev. Lett.* **110** 196802

Kim H-S *et al* 2012 Lead iodide perovskite sensitized all-solid-state submicron thin film mesoscopic solar cell with efficiency exceeding 9% *Sci. Rep.* **2** 591

Klem E J D, MacNeil D D, Cyr P W, Levina L and Sargent E H 2007 Efficient solution-processed infrared photovoltaic cells: Planarized all-inorganic bulk heterojunction devices via inter-quantum-dot bridging during growth from solution *Appl. Phys. Lett.* **90** 183113

Klimov V I 2017 *Nanocrystal Quantum Dots* 2nd edn (Boca Raton, FL: CRC Press)

Kojima A, Teshima K, Shirai Y and Miyasaka T 2009 Organometal halide perovskites as visible-light sensitizers for photovoltaic cells *J. Am. Chem. Soc.* **131** 6050–1

Konstantatos G and Sargent E H 2013 *Colloidal Quantum Dot Optoelectronics and Photovoltaics* (Cambridge: Cambridge University Press)

Kramer I J and Sargent E H 2014 The architecture of colloidal quantum dot solar cells: materials to devices *Chem. Rev.* **114** 863–82

Lan X, Masala S and Sargent E H 2014 Charge-extraction strategies for colloidal quantum dot photovoltaics *Nat. Mater.* **13** 233–40

Lan X *et al* 2016 10.6% certified colloidal quantum dot solar cells via solvent-polarity-engineered halide passivation *Nano Lett.* **16** 4630–4

Lee M M, Teuscher J, Miyasaka T, Murakami T N and Snaith H J 2012 Efficient hybrid solar cells based on meso-superstructured organometal halide perovskites *Science* **338** 643–7

Leijtens T, Bush K, Cheacharoen R, Beal R, Bowring A and McGehee M D 2017 Towards enabling stable lead halide perovskite solar cells; interplay between structural, environmental, and thermal stability *J. Mater. Chem.* A **5** 11483–500

Leijtens T, Bush K A, Prasanna R and McGehee M D 2018 Opportunities and challenges for tandem solar cells using metal halide perovskite semiconductors *Nat. Energy* **3** 828–38

Liao W *et al* 2016 Fabrication of efficient low-bandgap perovskite solar cells by combining formamidinium tin iodide with methylammonium lead iodide *J. Am. Chem. Soc.* **138** 12360–3

Liu M, Johnston M B and Snaith H J 2013 Efficient planar heterojunction perovskite solar cells by vapour deposition *Nature* **501** 395–8

Liu M *et al* 2017 Hybrid organic–inorganic inks flatten the energy landscape in colloidal quantum dot solids *Nat. Mater.* **16** 258–63

Liu Y, Gibbs M, Puthussery J, Gaik S, Ihly R, Hillhouse H W and Law M 2010 Dependence of carrier mobility on nanocrystal size and ligand length in PbSe nanocrystal solids *Nano Lett.* **10** 1960–9

Luque A and Hegedus S 2011 *Handbook of Photovoltaic Science and Engineering* (New York: Wiley)

Luther J M, Gao J, Lloyd M T, Semonin O E, Beard M C and Nozik A J 2010 Stability assessment on a 3% bilayer PbS/ZnO quantum dot heterojunction solar cell *Adv. Mater.* **22** 3704–7

Luther J M, Law M, Beard M C, Song Q, Reese M O, Ellingson R J and Nozik A J 2008 Schottky solar cells based on colloidal nanocrystal films *Nano Lett.* **8** 3488–92

Maria A, Cyr P W, Klem E J D, Levina L and Sargent E H 2005 Solution-processed infrared photovoltaic devices with >10% monochromatic internal quantum efficiency *Appl. Phys. Lett.* **87** 213112

McDonald S A, Konstantatos G, Zhang S, Cyr P W, Klem E J D, Levina L and Sargent E H 2005 Solution-processed PbS quantum dot infrared photodetectors and photovoltaics *Nat. Mater.* **4** 138–42

Mitzi D B 1999 Synthesis, structure, and properties of organic–inorganic perovskites and related materials *Progress in Inorganic Chemistry* (New York: Wiley) pp 1–121

Miyata A, Mitioglu A, Plochocka P, Portugall O, Wang J T-W, Stranks S D, Snaith H J and Nicholas R J 2015 Direct measurement of the exciton binding energy and effective masses for charge carriers in organic–inorganic tri-halide perovskites *Nat. Phys.* **11** 582–7

Murray C B, Kagan C R and Bawendi M G 2000 Synthesis and characterization of monodisperse nanocrystals and close-packed nanocrystal assemblies *Annu. Rev. Mater. Sci.* **30** 545–610

Murray C B, Norris D J and Bawendi M G 1993 Synthesis and characterization of nearly monodisperse CdE (E = sulfur, selenium, tellurium) semiconductor nanocrystallites *J. Am. Chem. Soc.* **115** 8706–15

National Renewable Energy Laboratory 2019 Best research-cell efficiency chart https://www.nrel.gov/pv/cell-efficiency.html

Nelson C A and Zhu X-Y 2012 Reversible surface electronic traps in PbS quantum dot solids induced by an order–disorder phase transition in capping molecules *J. Am. Chem. Soc.* **134** 7592–5

Nelson J 2003 *The Physics of Solar Cells* (Singapore: World Scientific)

Ning Z *et al* 2014 Air-stable n-type colloidal quantum dot solids *Nat. Mater.* **13** 822–8

Osedach T P, Zhao N, Andrew T L, Brown P R, Wanger D D, Strasfeld D B, Chang L-Y, Bawendi M G and Bulović V 2012 Bias-stress effect in 1,2-ethanedithiol-treated PbS quantum dot field-effect transistors *ACS Nano* **6** 3121–7

Park N-G, Grätzel M, Miyasaka T, Zhu K and Emery K 2016 Towards stable and commercially available perovskite solar cells *Nat. Energy* **1** 16152

Pattantyus-Abraham A G *et al* 2010 Depleted-heterojunction colloidal quantum dot solar cells *ACS Nano* **4** 3374–80

Pazos-Outón L M *et al* 2016 Photon recycling in lead iodide perovskite solar cells *Science* **351** 1430–3

Peters I M, Sofia S, Mailoa J and Buonassisi T 2016 Techno-economic analysis of tandem photovoltaic systems *RSC Adv.* **6** 66911–23

Petrus M L, Schlipf J, Li C, Gujar T P, Giesbrecht N, Müller-Buschbaum P, Thelakkat M, Bein T, Hüttner S and Docampo P 2017 Capturing the sun: a review of the challenges and perspectives of perovskite solar cells *Adv. Energy Mater.* **7** 1700264

Sanehira E M, Marshall A R, Christians J A, Harvey S P, Ciesielski P N, Wheeler L M, Schulz P, Lin L Y, Beard M C and Luther J M 2017 Enhanced mobility CsPbI$_3$ quantum dot arrays for record-efficiency, high-voltage photovoltaic cells *Sci. Adv.* **3** eaao4204

Sargent E H 2012 Colloidal quantum dot solar cells *Nat. Photonics* **6** 133

Shi D *et al* 2015 Low trap-state density and long carrier diffusion in organolead trihalide perovskite single crystals *Science* **347** 519–22

Snaith H J 2013 Perovskites: the emergence of a new era for low-cost, high-efficiency solar cells *J. Phys. Chem. Lett.* **4** 3623–30

Song Z, McElvany C L, Phillips A B, Celik I, Krantz P W, Watthage S C, Liyanage G K, Apul D and Heben M J 2017 A technoeconomic analysis of perovskite solar module manufacturing with low-cost materials and techniques *Energy Environ. Sci.* **10** 1297–305

Stranks S D, Eperon G E, Grancini G, Menelaou C, Alcocer M J P, Leijtens T, Herz L M, Petrozza A and Snaith H J 2013 Electron-hole diffusion lengths exceeding 1 micrometer in an organometal trihalide perovskite absorber *Science* **342** 341–4

Swarnkar A, Marshall A R, Sanehira E M, Chernomordik B D, Moore D T, Christians J A, Chakrabarti T and Luther J M 2016 Quantum dot-induced phase stabilization of α-CsPbI$_3$ perovskite for high-efficiency photovoltaics *Science* **354** 92–5

Talgorn E, Gao Y, Aerts M, Kunneman L T, Schins J M, Savenije T J, van Huis M A, van der Zant H S J, Houtepen A J and Siebbeles L D A 2011 Unity quantum yield of photogenerated charges and band-like transport in quantum-dot solids *Nat. Nanotechnol.* **6** 733–9

Tang J *et al* 2011 Colloidal-quantum-dot photovoltaics using atomic-ligand passivation *Nat. Mater.* **10** 765–71

Tang J and Sargent E H 2011 Infrared colloidal quantum dots for photovoltaics: fundamentals and recent progress *Adv. Mater.* **23** 12–29

Thompson N J *et al* 2014 Energy harvesting of non-emissive triplet excitons in tetracene by emissive PbS nanocrystals *Nat. Mater.* **13** 1039–43

Todorov T, Gunawan O and Guha S 2016 A road towards 25% efficiency and beyond: perovskite tandem solar cells *Mol. Syst. Des. Eng.* **1** 370–6

Tvingstedt K, Malinkiewicz O, Baumann A, Deibel C, Snaith H J, Dyakonov V and Bolink H J 2014 Radiative efficiency of lead iodide based perovskite solar cells *Sci. Rep.* **4** 6071

Wehrenfennig C, Eperon G E, Johnston M B, Snaith H J and Herz L M 2014 High charge carrier mobilities and lifetimes in organolead trihalide perovskites *Adv. Mater.* **26** 1584–9

Wise F W 2000 Lead salt quantum dots: the limit of strong quantum confinement *Acc. Chem. Res.* **33** 773–80

Würfel P and Würfel U 2016 *Physics of Solar Cells: From Basic Principles to Advanced Concepts* (New York: Wiley)

Yablonovitch E 2016 Lead halides join the top optoelectronic league *Science* **351** 1401

Yin W J, Shi T T and Yan Y F 2014 Unusual defect physics in $CH_3NH_3PbI_3$ perovskite solar cell absorber *Appl. Phys. Lett.* **104** 063903

Yuan M, Liu M and Sargent E H 2016 Colloidal quantum dot solids for solution-processed solar cells *Nat. Energy* **1** 16016

Zhang Y, Zherebetskyy D, Bronstein N D, Barja S, Lichtenstein L, Alivisatos A P, Wang L-W and Salmeron M 2015 Molecular oxygen induced in-gap states in PbS quantum dots *ACS Nano* **9** 10445–52

Chapter 5

R&D and commercialization challenges for emerging PV technologies

Because of the similarities discussed in the previous chapter, many important R&D challenges toward commercialization are shared between perovskite and quantum dot (QD) solar cells (figure 5.1). The most widely recognized challenge is that of reaching higher efficiencies than today's commercial technologies. Doing so will likely require the continued development of tandem cell architectures. Another major challenge involves scaling up solution and vacuum-based deposition processes to large areas while maintaining uniformity in terms of film thickness, film morphology, and device performance. These processes will ideally achieve high throughput to minimize capex and use low-cost, scalable precursors to minimize module cost. To enable flexible PV applications, further development of low-cost yet effective transparent moisture and oxygen barriers is critical.

One commonly cited problem is the presence of toxic lead in the most efficient, stable perovskite and QD materials to date. A 1 μm thick perovskite film contains roughly 1.5 g of lead per square meter, comparable to the amount of toxic cadmium in commercial CdTe solar cells. However, the bioavailability and environmental chemistry of lead in perovskite and quantum-dot materials requires further study; even if environmental and health risks are found to be minimal, public perception may dictate the commercial acceptance of any lead-based technology. In any case, the higher lead concentrations encountered during manufacturing and recycling/disposal must be managed with appropriate controls to protect workers and nearby communities. Lead-containing byproducts and modules may have to be disposed of as hazardous waste. Both of these considerations could increase lifecycle costs.

Early-stage materials R&D can help mitigate long-term environmental health and safety risks. For example, recent work shows that lead can be effectively sequestered in perovskite devices without sacrificing efficiency, avoiding lead leakage even when the device packaging is severely compromised (Li *et al* 2020). The proposed strategy involves sandwiching devices between a molecular film

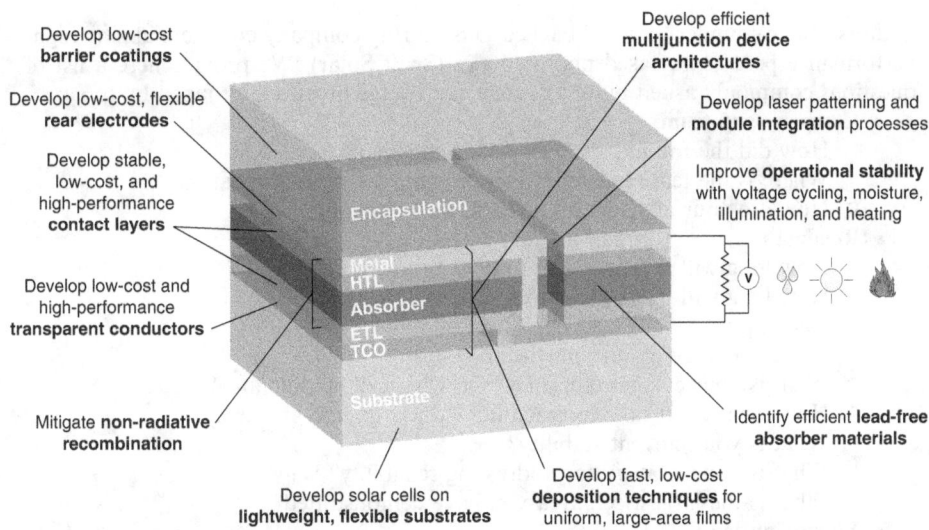

Figure 5.1. Shared R&D challenges for perovskite and QD solar cells. Many aspects of the materials, device structures, and manufacturing processes used in emerging PV technologies are important topics of investigation for academic and industrial R&D.

containing phosphonic acid groups and a lead-chelating polymer blend, which together absorb over 96% of the leached lead. The method used to prepare PbS QD inks can affect both the lead concentration in synthesis byproducts and the lead leaching potential from complete cells (Moody *et al* 2019). Ultimately, the development of recycling and other full life cycle management strategies will be important for large-scale deployment of any lead-based PV technology.

Building a new solar technology startup

The historical success rate of new PV technology startups worldwide has been abysmally low. Since around 2010, PV startups have struggled to scale up and compete with Chinese c-Si companies, which have rapidly expanded manufacturing capacity and driven down module prices, along with profit margins. This is true across many different PV technologies, including thin-film crystalline silicon, amorphous silicon, III–V multijunction, CdTe, CIGS, dye-sensitized, organic, and polymer solar cells.

Perovskites are the newest PV technology to enter the commercial arena. Perovskites stand out from other emerging technologies for many reasons described in the previous chapter. In short, they meet an unprecedented number of performance requirements for a competitive PV technology—high efficiency, high power-to-weight ratio, low material usage, high material production volumes, scalable manufacturing, and compatibility with lightweight and flexible substrates. The possibility of pairing perovskites in tandem with established technologies provides a second pathway to market that was not available for earlier technologies.

To build a new PV company, entrepreneurs must raise a substantial amount of capital, largely from venture capital (VC) investors and government grants. One of the authors is a co-founder of a VC-backed U.S. startup company commercializing high-performance perovskite-based photovoltaics (Swift Solar). We provide here a list of questions commonly asked by prospective early-stage investors, grouped by topic:

- **Company and team**
  - How did the founders start working together?
  - Where is the team located? Why did you choose that location?
  - How are your advisors involved in the company?
- **Roadmap**
  - How long will it take to get to pilot production?
  - How long will it take to make an initial product?
- **Technology**
  - What is unique about your technology?
  - What is your current efficiency? On what cell/module area?
  - How long will it take to reach 20%, 25%, or 30% efficiency?
  - What is your current stability?
  - What is your strategy for addressing reliability? Why?
  - What is the largest cell area you have demonstrated?
- **Manufacturing**
  - Do you plan to manufacture the technology yourself or work with a partner?
  - What manufacturing methods do you use? How fast can they go?
  - Can you use off-the-shelf tools?
  - What is the ultimate module cost per watt you expect to reach?
- **Intellectual property (IP)**
  - Is the technology protected by patents?
  - What IP do you own or have access to? What does it cover?
- **Market**
  - How do you plan to enter the market?
  - What are your initial markets? Are there other options?
  - Where do you envision future revenue coming from?
  - What are the unit economics?
  - How willing is the industry to accept shorter-lived modules?
  - Will lead-based materials be an issue?
- **Competition**
  - Who are your main competitors?
  - Which other companies globally are working on this technology?
  - What stops established companies from starting to work on your technology?
- **Financing**
  - How much funding have you raised to date? On what terms?
  - Have you received any grants?
  - What is your current burn rate?
  - Who are your current investors? Do you have any strategic investors?
  - How much money are you raising now? What will the funding be used for?
  - Who is leading this round? Who else is participating in the round?
  - What is your funding roadmap?
  - How much capital do you need to reach pilot or large-scale production?

5.1 The stability challenge

Perhaps the most important obstacle to commercialization and to realization of the ultimate potential of any emerging PV technology—aside from low efficiencies for QDs—is operational stability, or the consistency of device performance (efficiency) over time. For perovskites, extensive research efforts have led to rapid improvement in device-level stability, with a T80 lifetime of over 1000 h demonstrated for many different device architectures and test conditions. Similar demonstrations are still lacking for QD solar cells. For both technologies, much further work is needed to prove operational stability at the cell and module level and to reach a module lifetime of 15 or more years.

Over the last few years, perovskite stability studies have produced an improved understanding of the many factors that can influence either the perovskite layer alone or the full device stack[1]. One way to think about device stability is to break down these factors into mechanisms leading to intrinsic instability, which occur even under ideal PV operating conditions (i.e. light and electrical load), and sources of extrinsic stress, which may exacerbate device degradation.

The main source of intrinsic instability in perovskite solar cells is ion migration. Perovskites are mixed ionic-electronic conductors: Cations and anions can both move preferentially under the influence of an electric field (i.e. under normal solar cell operation), potentially leading to hysteresis and degradation of photovoltaic performance. For example, light-induced phase segregation is a well-known issue for mixed-anion compositions—the halides tend to segregate and form iodide- and bromide-rich regions upon light exposure, reducing the maximum photovoltage attainable (Hoke *et al* 2015). Even for single-anion compositions, migration of halides to a metal electrode can lead to detrimental reactions, such as the formation of silver or aluminum halide compounds (Kato *et al* 2015). The opposite process of metal migration from electrodes to the perovskite layer can also degrade device performance by creating shunts or deep trap states (Domanski *et al* 2016). Once cells are integrated into a module, interconnects can produce unique degradation mechanisms related to interactions between perovskites, metals, and contact/transport layers.

Extrinsic stressors such as moisture, oxygen, elevated temperatures, UV exposure, and mechanical stress—or a combination of these—can exacerbate device degradation if not appropriately managed. These stressors may be naturally present under practical operating conditions or intentionally introduced in the lab to accelerate the study of known aging processes. The effect of moisture on perovskite device performance is well-documented. Relatively small amounts of moisture can cause decomposition of perovskite absorbers; when exposed to high humidity, unencapsulated perovskite cells degrade rapidly (Domanski *et al* 2018). That said, controlled exposure to small amounts of moisture during film formation can improve the

---

[1] Degradation pathways in all of the device layers are often interrelated. For example, one recent study found that multiple contact layers and interfaces had to be jointly optimized to significantly improve device stability (Christians *et al* 2018).

perovskite optoelectronic quality and device performance (Eperon *et al* 2015a). Oxygen exposure can in some cases be as detrimental as moisture, although it is generally harmful only under illumination (i.e. photo-oxidation) (Aristidou *et al* 2015, Bryant *et al* 2016, Pearson *et al* 2016, Sheikh *et al* 2015). Elevated temperatures can cause thermal decomposition of perovskites—for example, MAPbI$_3$ begins degrading at relatively low temperatures (<100 °C) due to the volatility of methylammonium. At PV operating temperatures (–40 °C to 85 °C or higher), some perovskite compositions can undergo unfavorable phase changes that lead to device failure. Exposure to UV light can degrade TiO$_2$-based devices (Leijtens *et al* 2013) but is generally benign for perovskites under inert conditions (Domanski *et al* 2018). Mechanical stress may induce delamination and device failure in flexible applications, due to the low adhesion between perovskite grains and between layers in the device stack (Rolston *et al* 2016).

Many engineering strategies have been shown to improve perovskite stability in the face of these intrinsic and extrinsic stressors. Changing the perovskite composition is one leading approach—for example, various stability improvements have been achieved by partially replacing iodide with bromide, methylammonium with formamidinium (Eperon *et al* 2014, Jeon *et al* 2015), and organic cations with cesium (Beal *et al* 2016, Eperon *et al* 2015b, McMeekin *et al* 2016, Swarnkar *et al* 2016) and/or rubidium (Saliba *et al* 2016). Reducing perovskite dimensionality to form 2D layered structures—with layers of hydrophobic organic cations separating blocks of 3D perovskite—can improve stability significantly by protecting the 3D perovskite from moisture, although it also increases bandgap and exciton binding energy (figure 5.2) (Quan *et al* 2016, Smith *et al* 2014, Tsai *et al* 2016). Perovskite nanocrystals also appear to be more phase-stable than their bulk counterparts (Swarnkar *et al* 2016). Modifying the device structure is a proven approach to

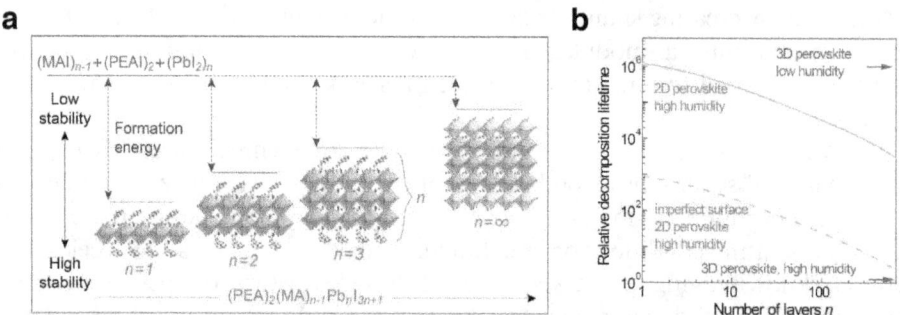

Figure 5.2. Stability enhancement with layered 2D perovskites. (a) Unit cell structure of (C$_8$H$_9$NH$_3$)$_2$(CH$_3$NH$_3$)$_{n-1}$Pb$_n$I$_{3n+1}$ perovskites with different numbers of layers (n), showing the evolution of dimensionality from 2D (n = 1) to 3D (n = ∞). (b) Density functional theory (DFT) calculations of the formation energy of perovskites with different n values in different atmospheres, showing that the van der Waals interactions between capping organic molecules in 2D perovskites significantly improve material stability in the presence of moisture. Reprinted with permission from Quan *et al* (2016). Copyright 2016 American Chemical Society.

mitigating known instabilities—for example, the use of moisture-blocking materials as transport layers (Habisreutinger *et al* 2016) or as electrodes (Bush *et al* 2016, Deng *et al* 2016) can dramatically improve extrinsic stability. Finally, using effective encapsulation methods can largely eliminate oxygen- and moisture-induced degradation.

An important ongoing research effort in the perovskite field involves the standardization of procedures for perovskite stability assessment and reporting. No such global standards exist today, although some have been proposed (Khenkin *et al* 2020). Developing universally applicable tests is difficult in part because degradation mechanisms vary widely for different perovskite compositions and device structures. Furthermore, due to ion migration, perovskite devices can take hours to reach steady-state conditions and can exhibit reversible performance loss—for example, over day/night cycling (Domanski *et al* 2017). As a result, tests that do not induce ion migration—such as shelf-life measurements (i.e. dark storage under ambient or elevated temperature)—are generally not representative of operational stability.

An effective stability test should imitate the illumination, electrical load, and environmental conditions (i.e. relative humidity and temperature) of typical solar cell operation. Evidence suggests that the most reliable long-term degradation test for perovskite solar cells is aging in an inert environment under 1 sun illumination with maximum power point (MPP) tracking, under both room and elevated temperatures (Domanski *et al* 2018, Saliba 2018).

Using encapsulation adds another variable to stability tests, as it is sometimes unclear what is being tested—the device's intrinsic stability or the quality and method of encapsulation, or both. Testing of encapsulated cells and modules is necessary to demonstrate product-level reliability. However, if the goal is to understand intrinsic stability or stability under one or more specific extrinsic stressors, cell testing should be carried out in controlled environments without encapsulation. For perovskites in particular, one potential downside of testing without encapsulation is that continuous flushing with gas may reduce the partial pressure of volatile decomposition products such as hydrogen iodide and methylamine, increasing decomposition rates. One proposed 'safe' solution is thus to age encapsulated devices in inert atmosphere (Domanski *et al* 2018).

Truly long-term PV performance tests are not feasible in the laboratory. A common lab test duration is 1000 h under peak sunlight, equivalent to 42 days of continuous peak operation or roughly 200 days of average solar insolation at 5 kWh/m^2/day. In contrast, commercial PV modules today are expected to perform with less than 20% cumulative degradation over 25 years, equivalent to roughly 9125 diurnal cycles or over 45 000 h of peak irradiance. To enable shorter yet informative tests, one can accelerate the aging process by selectively applying elevated light intensity, humidity, temperature cycling, or a combination of the above. The industry-standard module qualification tests (IEC 61215 for c-Si modules and IEC 61646 for thin-film modules) are accelerated stress tests that target specific failure mechanisms observed in the field for c-Si, CdTe, CIGS, and a-Si:H modules[2]. For these commercial technologies, the standard qualification tests have been validated extensively and have dramatically reduced module failure rates in the field. For perovskites and other emerging PV technologies, however, there may be new failure modes that are not stressed by the existing IEC tests. New accelerated aging tests must therefore be developed to ensure long-term field reliability.

5.2 Outlook for emerging PV technologies

Most directions for future perovskite and QD solar cell research can be categorized by their primary goal—either to improve our understanding of device-relevant phenomena or to improve solar cell performance, stability, and manufacturability by engineering materials, device architectures, and processing methods. For QDs, important questions remain about the origin of mid-gap recombination centers, techniques for mitigating gap states, and the effect of air exposure and environmental conditions on device operation. One critical technological challenge is the development of low-cost and scalable techniques for synthesizing colloidal QDs. For perovskites, important scientific questions revolve around the effect of multiple cations, defect tolerance, passivation, and ion migration. Important technological directions include surface and bulk passivation methods, lead-free perovskites (Giustino and Snaith 2016), and low-dimensional perovskites (e.g. 2-D layered perovskites (Tsai *et al* 2016) and perovskite QDs (Sanehira *et al* 2017, Swarnkar *et al* 2016)), which may be more stable than conventional polycrystalline materials.

In closing, it would be a grave mistake to not deploy the commercial PV technologies of today as we develop the technologies of tomorrow. Climate change leaves no time to spare. But the deployment of terawatts of solar worldwide by mid-century may be accelerated by the very-low cost, high performance, and versatile form factor of emerging PV technologies. By heeding the market forces that drive deployment, PV researchers can help pave the way for a ubiquitous solar future.

References

Aristidou N, Sanchez-Molina I, Chotchuangchutchaval T, Brown M, Martinez L, Rath T and Haque S A 2015 The role of oxygen in the degradation of methylammonium lead trihalide perovskite photoactive layers *Angew. Chem. Int. Ed. Engl.* **54** 8208–12

Beal R E, Slotcavage D J, Leijtens T, Bowring A R, Belisle R A, Nguyen W H, Burkhard G F, Hoke E T and McGehee M D 2016 Cesium lead halide perovskites with improved stability for tandem solar cells *J. Phys. Chem. Lett.* **7** 746–51

Bryant D, Aristidou N, Pont S, Sanchez-Molina I, Chotchunangatchaval T, Wheeler S, Durrant J R and Haque S A 2016 Light and oxygen induced degradation limits the operational stability of methylammonium lead triiodide perovskite solar cells *Energy Environ. Sci.* **9** 1655–60

---

[2] The IEC accelerated stress tests include thermal cycling (200 cycles from −40 °C to 90 °C with peak current), damp heat (1000 h at 85 °C and 85% relative humidity), humidity freeze (10 cycles from 85 °C to −40 °C at 85% relative humidity), mechanical load, hail, wet leakage current, hot spot, UV exposure, and bypass diode thermal tests.

Bush K A, Bailie C D, Chen Y, Bowring A R, Wang W, Ma W, Leijtens T, Moghadam F and McGehee M D 2016 Thermal and environmental stability of semi-transparent perovskite solar cells for tandems enabled by a solution-processed nanoparticle buffer layer and sputtered ITO electrode *Adv. Mater.* **28** 3937–43

Christians J A, Schulz P, Tinkham J S, Schloemer T H, Harvey S P, de Villers B J T, Sellinger A, Berry J J and Luther J M 2018 Tailored interfaces of unencapsulated perovskite solar cells for >1000 hour operational stability *Nat. Energy* **3** 68–74

Deng Y, Dong Q, Bi C, Yuan Y and Huang J 2016 Air-stable, efficient mixed-cation perovskite solar cells with Cu electrode by scalable fabrication of active layer *Adv. Energy Mater.* **6** 1600372

Domanski K 2017 Migration of cations induces reversible performance losses over day/night cycling in perovskite solar cells *Energy Environ. Sci.* **10** 604–13

Domanski K, Alharbi E A, Hagfeldt A, Grätzel M and Tress W 2018 Systematic investigation of the impact of operation conditions on the degradation behaviour of perovskite solar cells *Nat. Energy* **3** 61–7

Domanski K, Correa-Baena J-P, Mine N, Nazeeruddin M K, Abate A, Saliba M, Tress W, Hagfeldt A and Grätzel M 2016 Not all that glitters is gold: metal-migration-induced degradation in perovskite solar cells *ACS Nano* **10** 6306–14

Eperon G E *et al* 2015a The importance of moisture in hybrid lead halide perovskite thin film fabrication *ACS Nano* **9** 9380–93

Eperon G E, Paternò G M, Sutton R J, Zampetti A, Haghighirad A A, Cacialli F and Snaith H J 2015b Inorganic caesium lead iodide perovskite solar cells *J. Mater. Chem. A* **3** 19688–95

Eperon G E, Stranks S D, Menelaou C, Johnston M B, Herz L M and Snaith H J 2014 Formamidinium lead trihalide: a broadly tunable perovskite for efficient planar hetero-junction solar cells *Energy Environ. Sci.* **7** 982–8

Giustino F and Snaith H J 2016 Toward lead-free perovskite solar cells *ACS Energy Lett.* **1** 1233–40

Habisreutinger S N, McMeekin D P, Snaith H J and Nicholas R J 2016 Research Update: Strategies for improving the stability of perovskite solar cells *APL Mater.* **4** 091503

Hoke E T, Slotcavage D J, Dohner E R, Bowring A R, Karunadasa H I and McGehee M D 2015 Reversible photo-induced trap formation in mixed-halide hybrid perovskites for photovoltaics *Chem. Sci.* **6** 613–7

Jeon N J, Noh J H, Yang W S, Kim Y C, Ryu S, Seo J and Seok S I 2015 Compositional engineering of perovskite materials for high-performance solar cells *Nature* **517** 476–80

Kato Y, Ono L K, Lee M V, Wang S, Raga S R and Qi Y 2015 Silver iodide formation in methyl ammonium lead iodide perovskite solar cells with silver top electrodes *Adv. Mater. Interfaces* **2** 1500195

Khenkin M V *et al* 2020 Consensus statement for stability assessment and reporting for perovskite photovoltaics based on ISOS procedures *Nature Energy* **5** 35–49

Leijtens T, Eperon G E, Pathak S, Abate A, Lee M M and Snaith H J 2013 Overcoming ultraviolet light instability of sensitized TiO_2 with meso-superstructured organometal tri-halide perovskite solar cells *Nat. Commun.* **4** 2885

Li X, Zhang F, He H, Berry J J, Zhu K and Xu T 2020 On-device lead sequestration for perovskite solar cells *Nature* **578** 555–8

McMeekin D P *et al* 2016 A mixed-cation lead mixed-halide perovskite absorber for tandem solar cells *Science* **351** 151–5

Moody N, Yoon D, Johnson A, Wassweiler E, Nasilowski M, Bulović V and Bawendi M G 2019 Decreased synthesis costs and waste product toxicity for lead sulfide quantum dot ink photovoltaics *Adv. Sustain. Syst* **13** 1900061

Pearson A J, Eperon G E, Hopkinson P E, Habisreutinger S N, Wang J T-W, Snaith H J and Greenham N C 2016 Oxygen degradation in mesoporous $Al_2O_3/CH_3NH_3PbI_{3-x}Cl_x$ perovskite solar cells: kinetics and mechanisms *Adv. Energy Mater.* **6** 1600014

Quan L N *et al* 2016 Ligand-stabilized reduced-dimensionality perovskites *J. Am. Chem. Soc.* **138** 2649–55

Rolston N *et al* 2016 Mechanical integrity of solution-processed perovskite solar cells *Extreme Mech. Lett.* **9** 353–8

Saliba M 2018 Perovskite solar cells must come of age *Science* **359** 388–9

Saliba M *et al* 2016 Incorporation of rubidium cations into perovskite solar cells improves photovoltaic performance *Science* **354** 206–9

Sanehira E M, Marshall A R, Christians J A, Harvey S P, Ciesielski P N, Wheeler L M, Schulz P, Lin L Y, Beard M C and Luther J M 2017 Enhanced mobility $CsPbI_3$ quantum dot arrays for record-efficiency, high-voltage photovoltaic cells *Sci. Adv.* **3** eaao4204

Sheikh A D, Bera A, Haque M A, Rakhi R B, Gobbo S D, Alshareef H N and Wu T 2015 Atmospheric effects on the photovoltaic performance of hybrid perovskite solar cells *Sol. Energy Mater. Sol. Cells* **137** 6–14

Smith I C, Hoke E T, Solis-Ibarra D, McGehee M D and Karunadasa H I 2014 A layered hybrid perovskite solar-cell absorber with enhanced moisture stability *Angew. Chem. Int. Ed. Engl.* **53** 11232–5

Swarnkar A, Marshall A R, Sanehira E M, Chernomordik B D, Moore D T, Christians J A, Chakrabarti T and Luther J M 2016 Quantum dot-induced phase stabilization of α-$CsPbI_3$ perovskite for high-efficiency photovoltaics *Science* **354** 92–5

Tsai H *et al* 2016 High-efficiency two-dimensional Ruddlesden–Popper perovskite solar cells *Nature* **536** 312–6

Emerging Photovoltaic Technologies

Joel Jean and Patrick Richard Brown

Appendix A

Useful resources for solar, energy, and climate change

This appendix lists a wide range of resources that we have found useful in our work in solar PV research, energy policy and systems research, climate change advocacy, and PV commercialization.

A.1 Reports

- **Climate**
  - Intergovernmental Panel on Climate Change (IPCC) Assessment Reports (https://www.ipcc.ch/reports/)
    - Periodic reports summarizing latest knowledge on climate change causes, impacts, and response options
  - National Climate Assessment Reports (https://nca2018.globalchange.gov/)
    - Periodic reports summarizing current understanding of current and future climate impacts across the United States
- **Energy technology and systems**
  - US Energy Information Administration (EIA) Annual Energy Outlook (AEO) (https://www.eia.gov/outlooks/aeo/)
    - Annual report with long-term projections for the U.S. energy system
  - International Energy Agency (IEA) World Energy Outlook (WEO) (https://www.iea.org/topics/world-energy-outlook)
    - Annual report with long-term projections for the global energy system
  - Lazard Levelized Cost of Energy Analysis (https://www.lazard.com/perspective/lcoe2019)
    - Annual LCOE comparison for various electricity generation technologies

- ○ REN21 Renewables Global Status Report (https://www.ren21.net/reports/global-status-report/)
  - ■ Annual report on renewable energy markets and policy trends
- **Solar**
  - ○ International Technology Roadmap for Photovoltaic (ITRPV) (https://itrpv.vdma.org/)
    - ■ Annual report of technology trends in crystalline silicon photovoltaics
  - ○ Fraunhofer Institute for Solar Energy Systems (ISE) PV Report (https://www.ise.fraunhofer.de/en/publications/studies/photovoltaics-report.html)
    - ■ Periodic update on PV technology and market development
  - ○ Lawrence Berkeley National Lab (LBNL) Tracking the Sun (https://emp.lbl.gov/tracking-the-sun)
    - ■ Annual report on installed price trends for US residential and commercial PV systems
  - ○ LBNL Utility-Scale Solar (https://emp.lbl.gov/utility-scale-solar)
    - ■ Annual report on installed cost, price, and performance trends for US utility-scale solar projects
  - ○ NREL US Solar Photovoltaic System Cost Benchmark (https://www.nrel.gov/analysis/solar-cost-analysis.html)
    - ■ Annual report of benchmark installed costs including cost breakdowns for US solar PV systems
  - ○ NREL Solar Industry Update (https://www.energy.gov/eere/solar/quarterly-solar-industry-update)
    - ■ Annual overview of US and global solar industry

A.2 Educational resources

- PVeducation.org
  - ○ https://www.pveducation.org/
  - ○ Introduction to PV technology and systems focusing on crystalline silicon solar cells and modules, with many illustrations, images, and interactive animations
- Jenny Nelson *The Physics of Solar Cells* (book)
- Jenny Chase *Solar Power Finance without The Jargon* (book)
- Alexandra von Meier *Electric Power Systems: A Conceptual Introduction* (book)
- David MacKay *Sustainable Energy—Without the Hot Air* (book) https://www.withouthotair.com/
- Robert Jaffe and Washington Taylor *The Physics of Energy* (book)
- Antonio Luque and Steven Hegedus (eds) *Handbook of Photovoltaic Science and Engineering* (book)

A.3 Datasets

- NREL Annual Technology Baseline (ATB)
  - https://atb.nrel.gov/
  - Annually updated dataset on current and projected future cost and performance parameters for various electricity generation technologies
- NREL Best Research-Cell Efficiency Chart
  - https://www.nrel.gov/pv/cell-efficiency.html
  - Definitive chart of the highest confirmed power conversion efficiencies for research cells for a range of photovoltaic technologies since 1976
- NREL Champion Photovoltaic Module Efficiency Chart
  - https://www.nrel.gov/pv/module-efficiency.html
  - Chart of the highest confirmed power conversion efficiencies for champion modules for a range of photovoltaic technologies since 1988
- National Solar Radiation Database (NSRDB)
  - https://nsrdb.nrel.gov/
  - Detailed solar radiation and meteorological data at hourly and half-hourly time intervals and high geographical resolution across the United States
- NASA MERRA-2 dataset
  - https://disc.gsfc.nasa.gov/datasets/M2T1NXRAD_5.12.4/summary
  - Historical global solar radiation at hourly resolution across the globe
- Renewables Ninja
  - https://www.renewables.ninja/
  - User-friendly interface to NASA MERRA-2 dataset
- NC State University Database of State Incentives for Renewables & Efficiency (DSIRE)
  - https://www.dsireusa.org/
  - Comprehensive database of US federal- and state-level incentives and policies for renewable energy and energy efficiency
- PV Module Price Index
  - https://www.pvxchange.com/en
  - Monthly updated data on PV module prices on the European spot market

A.4 PV modeling tools

- NREL PV Watts
  - https://pvwatts.nrel.gov/
  - Online calculator for estimating energy production and cost of residential and commercial PV systems
- NREL System Advisor Model (SAM)
  - https://sam.nrel.gov/
  - Software for calculating energy output, LCOE, and other technical and financial parameters for various financing models and types of renewable energy systems, including solar PV, wind, geothermal, biomass, battery storage, and others

- PC1D
    - https://sourceforge.net/projects/pc1d/
    - 1-D device simulator widely used in the c-Si PV industry
- PV Lighthouse
    - https://www.pvlighthouse.com.au/
    - Collection of online calculators for simulating solar cell operation and PV system economics
- PVLIB
    - https://pypi.org/project/pvlib/
    - Python library for simulating the performance of PV systems, based on tools originally developed at Sandia National Laboratories
- SCAPS-1D (Solar Cell Capacitance Simulator)
    - http://scaps.elis.ugent.be/
    - 1-D PV device simulator originally developed for CIGS and CdTe solar cells
- Transfer Matrix Optical Modeling
    - http://web.stanford.edu/group/mcgehee/transfermatrix/index.html
    - MATLAB® and Python code for simulating optical propagation and absorption in thin-film solar cells, developed by the McGehee group at Stanford

www.ingramcontent.com/pod-product-compliance
Lightning Source LLC
Chambersburg PA
CBHW080944170526
45158CB00008B/2374